Model Test Paper *of* Seed Science and Technology

K. Vanangamudi
M. Kokila
S. Kavitha
K. Raja

Competition Tutor

Published by
Competition Tutor
(Imprint of Scientific Publishers, INDIA)
5 A, New Pali Road, P.O. Box 91
Jodhpur 342 001 (INDIA)
E-mail: info@scientificpub.com
Website :www.scientificpub.com

ISBN: 978-93-89832-00-6
eISBN: 978-93-89832-01-3

Printed in India

The Authors

Dr. K. Vanangamudi, Former Dean (Agriculture) and Seed Scientist, graduated B.Sc., (Ag.) in 1975 from Agricultural College and Research Institute, Tamil Nadu Agricultural University, Coimbatore; M.Sc., (Ag.) and Ph.D. in Seed Science and Technology, respectively in 1977 and 1981 from TNAU; and Post Doctoral Fellow (1988-1990) at Mississippi State University, USA. He joined Department of Seed Science and Technology, TNAU as Assistant Professor in 1982 and served the University for 32 years as Associate Professor (1986-1988) at Agricultural Research Station, Bhavanisagar; Professor (1992-2001) at Forest College & Research Institute, Mettupalayam; Professor and Head (2001-2003) of Department of Seed Science and Technology; Dean, Adhiparasakthi Agricultural College (2003-2006), Kalavai; and Dean (Agriculture), Agricultural College and Research Institute, TNAU, Coimbatore (2006-2009). He retired from service on August 31, 2013

He has undergone Post Doctoral Programme in Tree Seed Technology at Mississippi State University, USA from 1988-1990 and visited USA, Japan and Canada in connection with Forestry education and research under USAID-Winrock Internationa! program. He underwent training on e-learning and web based teaching at UC, Davis and Cornell University, USA during 2017.

Dr. Vanangamudi's significant contributions in Agricultural Education are 1. Starting B.Tech (Bioinformatics), B.Tech (Agricultural Information Technology) and BS (Agribusiness Management) degrees at AC&RI, Coimbatore, and B.Sc (Horti.) degree at AP Agricultural College; 2. He introduced e-education and online examination; 3. Under NAIP project on "Development of e-courses for B.Sc., (Ag) degree program as Co.PI, he developed e-learning materials for all the 50 courses and hosted in TNAU and ICAR websites; and 4. Video streaming of lecture presentation by the teachers in the classroom (video filming, editing and publishing).

He has prepared course materials for 90 courses for Industrial Training Institute (ITI) program in Agriculture, Horticulture, Agricultural Engineering, Animal husbandry, Poultry, Food processing under "Mdodular Employment Skills" and "Centre of Excellence" funded by National Instructional Media Institute, Ministry of Labour & Employment, GOI, Chennai, for the students and instructors of ITI all over India.

As a researcher, he was the Principal Investigator of 11 research projects funded by ICAR-NAIP, ICAR-World bank. ICFRE-World bank, Tamil Nadu Forest department, Ministry of Labour and Employment (GOI), CIDA-Gulpeh, Canada and Private agencies.

Dr. Vanangamudi has published 2 review papers, 35 research papers in International journals, 112 research papers in National journals, 29 popular articles in English and 53 in Tamil. He has written 40 books in English and 10 in Tamil. He has also written 11 technical bulletins in English and 29 in Tamil.

In recognition of his contribution in Agriculture Education, Research and Extension, he was awarded with Best PG teacher award in 1998 and Best scientist award in 1999 by TNAU; Best PG Teacher award 2000 by Madras Agricultural Students Union (MASU), Coimbatore; Best Seed Scientist award in 2003 by Hisar Agricultural University (HAU), Hisar; Tamil Nadu Scientist award in 2006 by Tamil Nadu Council for Science and Technology (TNCST), Chennai; Best Research Paper award in 2005 by Indian Society of Seed Technology (ISST), New Delhi and Bharat Ratna Dr. C. Subramaniam Best Teacher award in 2006 by Indian Council of Agriculture Research (ICAR), New Delhi.

He is a recognized guide of Bharathiar University, Coimbatore; Bharathidasan University, Trichy and Forest Research Institute, Dehradun.

He was a Research and Technical Consultant to Little's Orientals Balm and Pharmaceuticals Ltd., Chennai for seed priming, seed film coating and setting up of Seed Research Lab; to a farmer for 200 acres of farm land development; to a EU-KKID-Peace trust project for end term evaluation; and to a Kerala Forest Department project (World Bank) for developing training course syllabi and establishment of seed processing and testing units. He served as Senior Advisor and Consultant to Dr Mohan's Health care Products Ltd., Chennai.

Ms. M. Kokila is now working as Junior Research Fellow, Trinity Cultural Academy, Coimbatore. She graduated B.Sc., (Agri.) from Agricultural College and Research Institute (TNAU), Kudumiyanmalai, Pudukkottai in 2019.

She has participated in Personality Development Workshop held on 21st & 22nd January 2019 in AC & RI, Kudumiyanmalai and undergone a Industrial training under Industrial Tie-Up Programme with J. M. Industry, Pudukkottai, Tamil Nadu. She has undergone ten days internship with NGO (Vrutti). She has conducted training to the farmers on seedling treatment with biofertilizer and sugarcane trash composting during her UG program at KVK, Vamban. She also served as a National Service Scheme volunteer (2015-17) and committee leader in various college functions (2018-2019).

Dr. S. Kavitha graduated B.sc., (Agri.) in 1999, M.Sc., (Agri.) and Ph.D. in Seed Science and Technology, respectively in 2002 and 2005 from Agricultural College and Research Institute, Tamil Nadu Agricultural University, Coimbatore. She was awarded with Senior Research Fellowship for doctoral research. She joined as Assistant Professor in the Department of Seed Science and Technology in 2009 and worked as a programme coordinator, KVK, Ramanathapuram since April 2018. Now she is working in the Department of Plant Genetic Resources, TNAU, Coimbatore since November, 2019. She has got ten years of experience in teaching, research and extension.

She has authored three books, 23 book chapters and 22 popular articles and published 14 research articles in national and international journals. She has excellent research experience by heading eight research projects as Principal Investigator and 10 research projects as Co-Principal Investigator funded by various agencies.

She has attended two International conferences, two International workshops, 20 National trainings, two 21 days trainings, seven National seminars and five workshops. She also conducted and co-ordinated various On Farm trials (OFt), Front Line Demonstrations (FLD), Farmers Field School (FFS), Integrated Farming System (IFS), Entrepreneur Development Programme (EDP), 88 on campus trainings, 32 off campus trainings, 3 vocational trainings, 6 seminars, 26 exhibitions, three sponsored trainings and delivered 78 special lectures as resource person to farmers and extension functionaries. She organized one ICAR winter school.

She has guided two M.Sc. (Ag) students as chairperson. She was awarded with Dr.M. Balakumarasamy award for the best Ph.D. student in 2005, Thiru. V.K. Kumarasamy award and Coimbatore Seed Corporation prize for the best M.Sc., (Agri.) student in Seed Science and Technology in 2002 and also received best paper (oral) presentation and best poster presentation awards.

Dr. K. Raja, Associate Professor working in the Department of Seed Science and Technology, Tamil Nadu Agricultural University, Coimbatore. He has vast experience in seed production, teaching, research and extension. He has received the Best Seed Production Scientist Award and Best Paper Presentation Awards. With several international and national articles to his credit he has authored three books that have been well received by the peer community.

Preface

Indian agriculture has made tremendous growth after Independence. Food production has increased from 50 million tons in 1947 to 283 million tons in 2017-18. The landmark achievements in agriculture since 1960's are due to introduction of high yielding varieties, increased use of fertilizers and other inputs, expansion of irrigation facilities, effective extension efforts, improved agronomic practices, and besides all, the valuable contribution of Indian farmers.

However, the use of high quality seeds plays a pivotal role in crop production to achieve food security. The performance and efficacy of other inputs also depend on the use of quality seed. Good quality seed alone contributes for 15-20 per cent increase in the yield of crops.

Importance of seed in Indian Agriculture gave a great scope to design curriculum on Seed Science and Technology in undergraduate and post graduate levels at different State Agricultural Universities. Students and scholars showed a lot of interest in specializing Seed Science and Technology and look for good positions in both public and private organizations. To secure a good score in competitive exams, they have to excel well in the exams as well as personal interviews and group discussions.

To meet out the needs of the students and scholars, efforts are taken to compile the book on "Model Tests of Seed Science and Technology". This book will be a useful guide to the students and scholars, who are specializing in Seed Science and Technology.

February, 2020 **K. VANANGAMUDI**

CONTENTS

MODEL TEST 1

1. **Seed is a/an**
   (A) Matured ovule (B) Ripened nucellus
   (C) Immature ovule (D) 3/4th matured ovule
2. **One male gamete fuses with polar nuclei to form**
   (A) Endosperm nucleus (B) Embryo
   (C) Ovary (D) Integument
3. **Example of a crop which reserves food in endosperm is**
   (A) Pulses (B) Vegetables
   (C) Maize (D) Oilseeds
4. **Identify the incorrect statement regarding grain**
   (i) Grain should be physically and genetically pure
   (ii) It may be converted to seed
   (iii) It should be treated with any chemicals
   (iv) Need to satisfy minimum seed certification standards
   (A) (ii) and(iii) (B) (i) alone
   (C) (iii) and(iv) (D) All of these
5. **Match the following**
   (a) Certified seed - 1. 100% genetic purity
   (b) Soybean - 2. 99% genetic purity
   (c) Nucleus seed - 3. 97% physical purity
   (d) Foundation seed - 4. 98% genetic purity
   (A) 4, 3, 1, 2 (B) 3, 4, 1, 2
   (C) 2, 1, 4, 3 (D) 2, 4, 1, 3
6. **Safe moisture percentage of a seed is**
   (A) 8 – 9% (B) 9 – 10%
   (C) < 8% (D) > 13%
7. **Seed generation system depends on**
   (A) Seed multiplication ratio (B) Rate of genetic deterioration
   (C) Total seed demand (D) All of these
8. **Match the following**
   (a) Nucleus seed - 1. White tag
   (b) Breeder seed - 2. Azure blue
   (c) Foundation seed - 3. No specific tag
   (d) Certified seed - 4. Golden yellow
   (A) 2, 1, 4, 3 (B) 3, 1, 4, 2
   (C) 3, 4, 1, 2 (D) 4, 3, 1, 2

9. Seed generation (GS) system followed for cross pollinated crop/ often cross pollinated crops is

(i) 3 GS (ii) 4 GS
(iii) 5 GS (iv) 2 GS

(A) (ii) and(iii) (B) (i) and(ii)
(C) (i),(ii) and(iii) (D) Any of these

10. Special class of paddy seed permitted in seed generation system is

(A) CS → CS I → CS II → CS III
(B) CS → CS I → CS II
(C) CS → CS I → CS II → CS III → CS IV
(D) CS → CS I

11. Progeny of breeder seed is

(A) Nucleus seed (B) Foundation seed
(C) Certified seed (D) Registered seed

12. Identify the incorrect statements

(i) Progeny of foundation seed is breeder seed
(ii) Breeder seed is maintained by concerned breeder of sponsoring institute
(iii) CS and FS is maintained by certification agency
(iv) Breeder seed is controlled by maintenance breeding

(A) (ii) and(iii) (B) (ii),(iii) and(iv)
(C) (i) alone (D) (i) and(iv)

13. Identify the incorrect statement about TFL

(A) Certification is compulsory for both notified and released varieties
(B) Satisfy the minimum field standards
(C) Tested for physical purity and germination
(D) Seed inspector alone take sample

14. Ability of a seed to withstand in field conditions are called

(A) Seed germination (B) Seed vigour
(C) Seed viability (D) Both (A) and (B)

15. Example of inseparable crop seed for wheat is

(A) Barley (B) Oats
(C) Gram (D) All of these

16. Incorrect statement about advantages of quality seed

(A) Yield increase by 15-20%
(B) Fetches higher price
(C) Genetically and physically pure
(D) Maximum replanting value

17. Fertilized and matured ovule is

(A) Ovary (B) Fruit
(C) Seedling (D) Seed

18. ***Argemone mexicana* is a inseparable crop seed in**
    (A) Wheat (B) Rape and mustard
    (C) Rice (D) Cotton

19. **Planting value of seed is**
    (A) Pure seed % - Germination %
    (B) Pure seed % x Germination %
    (C) Pure seed % / Germination %
    (D) None of these

20. **ISTA stands for**
    (A) International Seed Testing Agency
    (B) International Seed Testing Agreement
    (C) International Seed Testing Association
    (D) International Seed Testing Agenda

21. **Sowing report should be submitted within _____ days from the date of transplanting**
    (A) 30 days (B) 15 days
    (C) 45 days (D) 7 days

22. **Seed certification includes**
    (i) Field inspection
    (ii) Seed sampling
    (iii) Seed quality test
    (iv) Verification of seed source
    (A) (i),(ii) and (iv) are correct (B) (iii) and (iv) are correct
    (C) (i) alone (D) All of these

23. **Separate form of application is submitted in case of**
    (i) Different varieties
    (ii) Area more than 25 acres
    (iii) Date of sowing differs by 7 days or more
    (A) (i) and (ii) alone (B) (i) alone
    (C) (ii) and (iii) alone (D) All of these

24. **Detasseling is done in**
    (A) Sorghum (B) Cotton
    (C) Sugarcane (D) Maize

25. **Tag colour of registered seed is**
    (A) Opal green (B) Purple colour
    (C) White (D) Azure blue

26. **Seed which is produced by private seed companies is**
    (A) TFL seeds (B) Breeder seed
    (C) Nucleus seed (D) Both (A) and (B)

27. **Male sterility system followed for redgram is**
    (A) CGMS (B) TGMS
    (C) PGMS (D) GMS

28. Seed which is applicable to notified kinds only

(A) Certified seed (B) TFL seeds
(C) Foundation seed (D) Both (A) and (B)

29. Seed certification has

(A) 3 phases (B) 6 phases
(C) 7 phases (D) 4 phases

30. Father of seed testing is

(A) N. A. Cobb (B) Frederich Nobbe
(C) Dr. Badrinarayanan (D) Hiltner

31. Match the following

	Field size	*Number of counts*
(a)	0 – 5 acres	- 6 counts
(b)	11 – 15 acres	- 5 counts
(c)	16 – 20 acres	- 8 counts
(d)	6 – 10 acres	- 7 counts

(A) 3 4 1 2 (B) 2 3 4 1
(C) 2 1 4 3 (D) 2 4 3 1

32. Identify the correct statement

(A) During first inspection, source of the seed is not necessary to be verified
(B) Sowing report should be sent in duplicate form
(C) Presence of B line in A line is pollen partials
(D) Example of pollen shedders is rice

33. Maximum number of inspection is

(A) 3 (B) 2
(C) 4 (D) None of these

34. Match the following

(a) Shedding tassel - 1. Pearl millet
(b) Pollen partials - 2. Cotton
(c) Selfed bolls - 3. Rice
(d) Pollen shedders - 4. Maize

(A) 3 1 4 2 (B) 2 3 1 4
(C) 4 1 2 3 (D) 4 2 1 3

35. Number of plants/count taken for cotton is

(A) 500 (B) 1000
(C) 100 (D) None of these

36. Number of plants/earheads counted for tillering crops like rice is

(A) 500 (B) 1000
(C) 100 (D) None of these

37. Number of inspections followed for hybrids is

(A) 2 (B) 3
(C) 4 (D) Either (A) or (B)

38. **Match the following**

(a)	Widely spaced crop	- 1. Germination
(b)	Field standard	- 2. 100 plants
(c)	Seed standard	- 3. 1000 plants
(d)	No. of plants/inspection for broadcasted seeds	- 4. Isolation distance

(A) 1, 3, 2, 4 (B) 4, 3, 2, 1
(C) 3, 1, 4, 2 (D) 2, 4, 1, 3

39. **The validity period of certified seed is**
(A) 6 months (B) 9 months
(C) 10 months (D) 12 months

40. **Identify the incorrect statement**
(A) A small portion taken from representative sample is called working sample
(B) Submitted sample size is about 1 kg
(C) The working sample size of paddy is 40 g
(D) Nobbe triers is suitable for sampling a lot in bag

41. **Indented cylinder separator grades the seeds according to the**
(A) Weight (B) Surface characteristics
(C) Length (D) Size

42. **Match the following**

(a)	Specific gravity separator	- 1. Seed size
(b)	Magnetic separator	- 2. Seed weight
(c)	Spiral separator	- 3. Surface texture
(d)	Air screen cleaner	- 4. Seed shape

(A) 1 4 3 2 (B) 4 3 2 1
(C) 2 3 4 1 (D) 3 4 1 2

43. **The minimum germination percentage of paddy is**
(A) 60% (B) 75%
(C) 80% (D) 90%

44. **The minimum number of seeds used for germination test is**
(A) 250 (B) 400
(C) 1000 (D) None of these

45. **A specified quantity of seed which is physically identifiable to give certificate is**
(A) Seed certification (B) Seed lot
(C) Seed analysis (D) Seed testing

46. **Which of the following crop has maximum seed lot size?**
(A) Paddy (B) Groundnut
(C) Pearl millet (D) Maize

47. **_____ is to estimate the extent of biotic and abiotic impurities present in seed lot**
(A) Germination test (B) Purity analysis
(C) Standard test (D) None of these

48. Most practical and basic common method of moisture estimation prescribed by ISTA is

(A) Moisture meter (B) Hot air oven method
(C) Toluene method (D) Lyophilization

49. Purity analysis for groundnut is done with

(A) Kernels (B) Pods
(C) Nuts (D) Shell

50. RH in STL should be

(A) Less than 50% (B) Less than 40%
(C) Less than 70% (D) Less than 90%

51. Which of the following is an indirect method of moisture determination?

(A) Hot air oven method (B) Moisture meter method
(C) Lyophilization (D) Toluene method

52. Chemicals used in desiccators is/are

(A) $CaCl_2$ (B) Silica gel
(C) NaOH (D) All of these

53. Temperature range for high constant oven method is

(A) 103 ± 2°C (B) 130 ± 3°C
(C) 196 ± 2°C (D) 198 ± 3°C

54. Most commonly used paper for germination is/are

(i) Germination sheet (ii) Paddy straw paper
(iii) Filter paper (iv) Banana – pseudostem paper

(A) (i) and (iii) alone (B) (i) alone
(C) (i),(ii) and(iii) (D) All of these

55. Match the following

Crop	*Isolation distance*
(A) Rice hybrid	- 1. 800 m
(B) Maize	- 2. 100 m
(C) Bhendi(FS)	- 3. 200 m
(D) Pumpkin(FS)	- 4. 400 m

(A) 4 3 1 2 (B) 3 2 1 4
(C) 3 4 2 1 (D) 2 3 4 1

56. Germination test method for larger seeds is

(A) Top of paper (B) Between paper towel method
(C) Sand medium (D) All of these

57. Tetrazolium concentration used in viability test is

(A) 2% (B) 5%
(C) 0.5% (D) 10%

58. Grow out test is to predict the

(A) Physical purity of seed (B) Genetic purity of seed
(C) Seed viability (D) All of these

59. Example of seed viability test
(A) Accelerated ageing test (B) Brick gravel test
(C) Quick viability test (D) Paper piercing test

60. Seed vigour of cotton can be estimated by
(A) Cool germination test (B) Cold test
(C) Water sensitivity (D) Exhaustion test

61. On staining, the colour formed during quick viability test is
(A) Blue (B) Red
(C) Colourless (D) Orange

62. National Seed Corporation was set up in
(A) April 7, 1963 (B) March 7, 1963
(C) June 7, 1963 (D) January 7, 1963

63. Quick viability test was developed by
(A) Hiltner (B) Lakon
(C) Frederich Nobbe (D) None of these

64. (i) The concentration of $FeCl_3$ used for viability testing is 20%
(ii) Indoxyl acetate test is a specific seed viability test for soybean
(A) (i) alone is correct (B) (ii) alone is correct
(C) Both are correct (D) Both are wrong

65. Tag size of certified seed is
(A) 12 × 6 cm (B) 15 × 7.5 cm
(C) 13 × 6.5 cm (D) Any of these

66. Chemical used in accelerated ageing test is
(A) KOH (B) $FeCl_3$
(C) H_2SO_4 (D) Tetrazolium

67. Match the following

Seed vigour test	*Examples*
(A) Physical test	- 1. Cold test
(B) Biochemical test	- 2. Seed density
(C) Stress test	- 3. GADA test
(D) Performance test	- 4. Speed of germination

(A) 1, 3, 4, 2 (B) 2, 3, 1, 4
(C) 4, 3, 2, 1 (D) 2, 1, 4, 3

68. ISTA was formed during the year
(A) 1920 (B) 1924
(C) 1932 (D) 1928

69. (i) International seed lot certificate colour is orange
(ii) Headquaters of ISTA located in Switzerland
(iii) Number of countries registered in ISTA is 70
(A) (i) and(iii) are correct (B) (ii) and(iii) are correct
(C) (i) and(ii) are correct (D) All are correct

70. The first seed testing lab was established in 1869 is
(A) India (B) Europe
(C) Germany (D) Africa

71. The journal entitled "Seed Science and Technology" is published by ISTA since

(A) 1925 (B) 1956
(C) 1973 (D) 1960

72. International Seed Testing Congress was held every

(A) One year (B) Three years
(C) Five years (D) Ten years

73. Seed village concept is also known as

(A) Seed growers approach (B) Community area approach
(C) Farmer's seed approach (D) Compact area approach

74. Seed multiplication ratio is high for

(A) Paddy (B) Groundnut
(C) Finger millet (D) Pulses

75. The quality seeds which replaces the farm saved seeds is termed as

(A) Seed multiplication ratio (B) Seed vigour index
(C) Seed germination percentage (D) Seed replacement rate

76. Match the following

(A) Seed Rules – 1. 1966
(B) Seed Act – 2. 1983
(C) Seed Control Order – 3. 1968
(D) Essential Commodities Act – 4. 1955

(A) 3 1 2 4 (B) 4 3 1 2
(C) 1 3 2 4 (D) 2 3 4 1

77. Seed Act was implemented in the year

(A) 1967 (B) 1969
(C) 1966 (D) 1983

78. Notification of varieties comes under _____ of Seed Act.

(A) Sec 3 (B) Sec 5
(C) Sec 8 (D) Sec 11

79. Seed certification is for

(A) Breeder seed and certified seed
(B) Certified seed and foundation seed
(C) Certified seed alone
(D) All the above

80. Seed replacement rate is 100% for

(A) Varieties of self pollinated crops
(B) Varieties of cross pollinated crops
(C) Hybrids
(D) All of these

81. The suitable chemical used for seed hardening of paddy is

(A) HCl (1%) (B) KOH (1%)
(C) KH_2PO_4 (2%) (D) KCl (1%)

82. Match the following

	Sections of Seed Act	*Section*
(a)	Seed Analyst	– 1. Sec 4
(b)	Seed Inspector	– 2. Sec 12
(c)	Certification Agency	– 3. Sec 13
(d)	State Seed Laboratory	– 4. Sec 8

	a	b	c	d
A.	2	1	3	4
B.	3	2	4	1
C.	2	3	4	1
D.	4	3	2	1

83. Controlled hydration of seeds is called
(A) Seed hardening (B) Seed priming
(C) Seed encapsulation (D) Seed pelleting

84. Commonly used priming osmotium is
(A) Polyethylene glycol (B) Vermiculite
(C) Salt solution (D) *Pseudomonas*

85. New seed policy was evolved in
(A) 1966 (B) 1963
(C) 1988 (D) 1972

86. Central Seed Testing Laboratory only for testing BT cotton is located at
(A) Varanasi (B) Nagpur
(C) New Delhi (D) Bhopal

87. Number of Seed Testing Laboratory in Tamil Nadu is
(A) 7 (B) 10
(C) 8 (D) 12

88. Choose the correct statements
(i) Orthodox seed can tolerate low temperature
(ii) Racalcitrant seed can be dried below 30-35% moisture
(iii) Example of recalcitrant seeds is rubber and coffee
(iv) Orthodox seed can be stored for several years
(A) (i) and(ii) are correct
(B) (i) and(iii) are correct
(C) (i),(iii) and(iv) are correct
(D) All are correct

89. Seed borne fungi can be detected by
(A) Blotter method (B) Agar plate method
(C) NaOH seed soak test (D) All the above

90. Chemical which is not used to create male sterility is
(A) Maleic hydrazide (B) Ethrel
(C) Sodium methyl arsenate (D) NaOH

91. Commonly used chemical for encapsulation is

(A) Na alginate (B) NaCl
(C) Sodium pectate (D) NaOH

92. Stratification indicates the seed dormancy breaking method with

(A) High temperature
(B) Low temperature
(C) Initially low, later high temperature
(D) Ultra high temperature

93. Match the following

(a) Inoculant pelleting – 1. Napthalic anhydride
(b) Protective pelleting – 2. *Rhizobium*
(c) Herbicide pelleting – 3. *Trichoderma*
(d) Nutrient pelleting – 4. DAP

	(a)	(b)	(c)	(d)
(A)	3	2	4	1
(B)	2	3	1	4
(C)	2	1	4	3
(D)	4	3	1	2

94. Choose the correct statement

(i) Seed pelleting is done to facilitate precision planting
(ii) Seed colouring is to improve its brand identity and marketing
(iii) Chemical used for seed hardening of pulses is ZnSO4

(A) (i) and (ii) are correct
(B) (ii) and (iii) are correct
(C) (i) and (iii) are correct
(D) All are correct

95. National Seed Research and Training Centre is located at

(A) Coimbatore (B) Varanasi
(C) Nagpur (D) New Delhi

96. Find out the incorrect statement

(A) *In situ* conservation is known as conservation in natural habitats
(B) Example of *ex situ* conservation is gene bank.
(C) Millennium seed bank lab is located at London.
(D) Global seed vault is located at Germany

97. Cryopreservation was proposed for the first time in the year

(A) 1968 (B) 1978
(C) 1988 (D) 1962

98. Match the following

Invitro gene bank	***Place***
(a) CIMMYT, Mexico	1. Paddy
(b) CIP, Peru	2. Wheat and maize
(c) CIAT, Columbia	3. Potato
(e) IRRI, Phillipines	4. Cassava

(A) 2, 3, 4, 1 (B) 4, 3, 2, 1
(C) 1, 2, 4, 3 (D) 2, 4, 3, 1

99. Match the following

Spacing	***Crop***
(a) 20×30 cm	1. Cotton
(b) 45×30 cm	2. Sorghum
(c) 45×15 cm	3. Redgram
(3) 90×60 cm	4. Rice

(A) 4, 2, 3, 1 (B) 2, 1, 4, 3
(C) 3, 2, 1, 4 (D) 4, 3, 2, 1

100. Match the following

Crop	***Spacing***
(a) Maize	1. 30 × 15 cm
(b) Soybean	2. 40 × 20 cm
(c) Groundnut	3. 60 × 20 cm
(d) Safflower	4. 30 × 10 cm

(A) 4, 2, 3, 1 (B) 2, 1, 4, 3
(C) 3, 4, 1, 2 (D) 4, 3, 2, 1

Answer Keys

Q.No.	Ans.	Q.No.	Ans.	Q.No.	Ans.	Q.No.	Ans.	Q.No.	Ans.
1	A	21	B	41	C	61	B	81	D
2	A	22	D	42	C	62	B	82	C
3	C	23	D	43	C	63	B	83	B
4	D	24	D	44	B	64	C	84	A
5	A	25	B	45	B	65	B	85	C
6	B	26	A	46	D	66	A	86	B
7	D	27	D	47	B	67	B	87	B
8	C	28	A	48	B	68	B	88	C
9	B	29	B	49	B	69	C	89	D
10	B	30	B	50	C	70	C	90	D
11	B	31	D	51	B	71	C	91	A
12	D	32	D	52	D	72	B	92	B
13	B	33	C	53	B	73	D	93	B
14	B	34	C	54	C	74	C	94	D
15	D	35	C	55	D	75	D	95	B
16	D	36	B	56	C	76	A	96	D
17	D	37	C	57	C	77	B	97	A
18	B	38	D	58	B	78	B	98	A
19	B	39	B	59	C	79	B	99	D
20	C	40	A	60	A	80	C	100	C

MODEL TEST 2

1. **Hypanthodium is a characteristic inflorescence of**
(A) Mulberry
(B) Ficus
(C) Poinsettia
(D) Pine apple

2. **A flower is said to be complete when it has**
(A) Corolla
(B) Androecium and Gynoecium
(C) Calyx and Corolla only
(D) All the four whorls

3. **The fruit of gramineae is**
(A) Caruncle
(B) Caryopsis
(C) Achene
(D) Nut

4. **A fruit is best defined as**
(A) Ripened ovule enclosing one or more seeds
(B) Ripened ovary enclosing one or more seeds
(C) Ripened endosperm enclosing one or more seeds
(D) Ripened nucellus enclosing one or more seeds

5. **Nucellar polyembryony is reported in species of**
(A) Brassica
(B) Citrus
(C) Gossypium
(D) Triticum

6. **In apomixis, the progeny closely resembles :**
(A) Male parent
(B) Female parent
(C) Both male and female parents
(D) Restorer parent

7. **The storage protein in rice is**
(A) Glutelin
(B) Prolamin
(C) Globulin
(D) Albumin

8. **________ are the major form of phosphorous storage in seeds**
(A) Alkaloids
(B) Tannins
(C) Saponins
(D) Phytins

9. **An organic substance that can withstand environmental extremes can not be degraded by any enzyme is**
(A) Lignin
(B) Cellulose
(C) Cuticle
(D) Sporopollenin

10. **In which crop, the natural crossing is the major source of genetic contamination**
(A) Self fertilized crops
(B) Cross fertilized crops
(C) Self compatible crops
(D) Self incompatible crops

11. **Which one of the following is not dioecious plant?**
(A) Date palm
(B) Vallisneria
(C) Maize
(D) Papaya

12. The commonest form of the ovule position is
(A) Orthotropous (B) Anatropous
(C) Amphilotropous (D) Campylotropous

13. The main body of ovule is called
(A) Hilum (B) Chalaza
(C) Nucellus (D) Integuments

14. The tissue which nourish microspores is
(A) Nucellus (B) Mesophyll
(C) Tapetum (D) Endodermis

15. _______ is a plant that produces male flowers only, but does not produce seed.
(A) Monoecious (B) Dioecious
(C) Androecious (D) Gynoecious

16. Which type of mechanism favors cross pollination in anthurium?
(A) Protandry B) Protogyny
(C) Self incompatible (D) Insects

17. What is the optimum moisture for storage of marigold?
(A) 7% (B) 6%
(C) 9% (D) 8%

18. Which one of the following flowering plants possess dormancy mechanism?
(A) Zinnia (B) Gaillardia
(C) Cactus (D) Cosmos

19. A pure seed(minimum) of 97% is recommended for CS production of _______
(A) Annual chrysanthemum (B) Aster
(C) Ornamental sunflower (D) Marigold

20. Dry indehiscent single-seeded fruit formed from bicarpellary syncarpous inferior ovary is
(A) Cremocarp (B) Caryopsis
(C) Cypsela (D) Berry

21. Fruits are mature
(A) Ovules (B) Peduncles
(C) Ovaries (D) Flowers

22. Bolting refers to emergence of inflorescence in
(A) Sugarcane (B) Sugarbeet
(C) Potato (D) Sweet potato

23. Endosperm in gymnosperm is formed
(A) At the time of fertilization
(B) Before fertilization
(C) After fertilization
(D) Along with the development of embryo

24. Largest sperms are found in
(A) Pinus (B) Cedrus
(C) Cycas (D) Gnetum

25. Gymnosperm differ from angiosperm
(A) Having seeds (B) Having fruits
(C) Having naked ovules (D) None of these

26. Jack fruit seed is very short lived because it is ________
(A) Orthodox seed (B) Difficult recalcitrant seed
(C) Recalcitrant seed (D) Intermediate seed

27. Which of the following species does not possess dormancy?
(A) Loquat (B) Custard apple
(C) Pineapple (D) Litchi

28. The ________ dormancy is noticed in passion fruit seeds
(A) Endogenous dormancy (B) Chemical dormancy
(C) Seed coat dormancy (D) Embryo dormancy

29. What is the isolation distance for FS seed production of papaya?
(A) 3m (B) 4m
(C) 1m (D) 2m

30. The other living plants including rootstock for CS of guava is
(A) 1% (B) 2%
(C) 3% (D) 4%

31. The hormone which is involved in regulation of florigen synthesis
(A) Ethylene (B) Zeatin
(C) Cytokinin (D) Gibberellin

32. All are natural auxins except
(A) IAA (B) Phenoxy acetic acid
(C) Indole-3-acetic acid (D) NAA

33. Which of the following is an artificial ripening agent
(A) Ethephon (B) 2,4-D
(C) NAA (D) IAA

34. An example for chemical hybridizing agent is
(A) PEG (B) Mannitol
(C) MH (D) Calcium chloride

35. In which of the following crop, the CMS is employed for hybrid seed production
(A) Rice (B) Redgram
(C) Maize (D) Cotton

36. In which crop, rope pulling is done for effective pollination?
(A) Sunflower (B) Maize
(C) Rice (D) Pearl millet

37. To produce hybrid seed we need
(A) A male sterile line, restorer and maintainer
(B) A male fertile line and maintainer
(C) A maintainer and restorer
(D) Only male sterile line

38. Emasculation is
(A) Removal of male parts from a bisexual flower
(B) Removal of female plants
(C) Removal of male and female parts
(D) Removal of ovary

39. Which from the following is a male line of a three line hybrid seed production?
(A) A line (B) B line
(C) R line (D) M line

40. The production of double cross hybrid in maize was first suggested by
(A) Shull (B) East
(C) Jones (D) Hull

41. Pure line breed refers to
(A) Heterozygosity only
(B) Homozygostiy only
(C) Homozygosity and self assortment
(D) Heterozygosity and linkage

42. Which one of the following is not related to embryo culture?
(A) Used in propagation of haploid plants
(B) Embryos are used as explants
(C) Both immature and mature embryos are used
(D) To generate callus cultures or somatic embryos

43. The isolation distance of 400m is adopted for certified seed production of ___________
(A) Brinjal (B) Cucurbits
(C) Chilli (D) Onion mother bulb

44. What seed treatment is recommended for cucurbits?
(A) Use of pregerminated seed for sowing
(B) Hydropriming (36 h)
(C) Hydropriming (12 h)
(D) Sand matric 80% priming (24h)

45. What is the seed extraction for okra?
(A) Wet seed extraction (B) Dry seed extraction
(C) Acid / Alkali method (D) Manual extraction

46. Physical purity of 95% is permissible for the foundation and certified seed of
(A) Soybean (B) Groundnut
(C) Spinach (D) Carrot

47. The seeds of ________ is extracted by beating.

(A) Rice (B) Greengram
(C) Groundnut (D) Tomato

48. Which one of the following is the washing soda that is used in alkali method of seed extraction?

(A) Sodium carbonate (B) Sodium bicarbonate
(C) Sodium chloride (D) Sodium hydroxide

49. Elimination of moisture from the seed depends on ________ and ________ of the environment surrounding the seed

(A) RH and gases (B) RH and light
(C) RH and temperature (D) RH and rainfall

50. For drying of paddy seeds, the suitable drying method is ________

(A) Non mixing column driers
(B) Rectangular metal bin driers
(C) Circular metal bin driers
(D) Square metal bin driers

51. Which from the following machine is engaged for upgrading the seed based on weight?

(A) Air screen cleaner cum grader
(B) Indented cylinder separator
(C) Specific gravity separator
(D) Magnetic separator

52. Which one of following organic solvent is used for seed infusion?

(A) Ketone (B) Acetone
(C) Acetyl CoA (D) Acetic acid

53. What is the suggested temperature during seed priming?

(A) 0-5°C (B) 5-10°C
(C) 10-15°C (D) 15-20°C

54. For blackgram seed fortification, ________ is advocated

(A) 1% $CaCl_2$
(B) 2% KCl
(C) 0.5% Potassium dihydrogen chloride
(D) 100 ppm $ZnSO_4$

55. For egg floatation technique to upgrade paddy seeds, 2kg of commercial common salt is to be dissolved in ________ litres of water.

(A) 5 (B) 10
(C) 15 (D) 20

56. The designated disease of ________ is loose smut

(A) Sunflower (B) Rice
(C) Wheat (D) Cotton

57. After acid delinting and washing, which chemical is used to test the complete removal of acid
(A) Calcium oxychloride @ 0.5%
(B) Calcium bicarbonate @ 0.5%
(C) Calcium chloride @ 0.5%
(D) Calcium hydroxide @ 0.5%

58. Non-synchronous flowering is observed in the hybrid seed production of
(A) Pearl millet (B) Sorghum
(C) Redgram (D) Sunflower

59. Early seed plants were pollinated by
(A) Bees (B) Birds
(C) Wind (D) Butterflies

60. Which of the following is incorrect with respect to mode of pollination
(i) Rice-Wheat-Sorghum-Finger millet – Self pollinated
(ii) Maize-Pearl millet-Oil palm – Cross pollinated
(iii) Coconut-Mesta-Cotton pollinated – Often cross pollinated
(d) Garlic – No pollination
(A) (i) alone (B) (i) and (ii)
(C) (ii) and (iii) (D) (i) and (iii)

61. Germ pore is the region where the exine is
(A) Thin (B) Uniform
(C) Thin and uniform (D) Absent

62. Cotton
(A) Is a self-pollinated crop
(B) Is a cross pollinated crop
(C) Is an often-cross pollinated crop
(D) Is an often-self pollinated crop

63. What is the exact stage of removal of off types in seed field?
(A) After pollination (B) During fertilization
(C) Before pollination (D) After fertilization

64. Which is not relevant for over or excess irrigation of seed crops?
(A) Promote vegetative growth (B) Promote lodging
(C) Promote nutrient imbalance (D) Delayed flowering

65. In cotton, spraying of ________ DAP at boll formation increases seed quality
(A) 0.5% (B) 1%
(C) 1.5% (D) 2%

66. Which one of the following is not relevant for achieving synchronized flowering?
(A) Staggered sowing (B) Application of CCC
(C) Withholding irrigation (D) Application of urea or DAP

67. Nucellus in seed is represented by
(A) Testa (B) Pericarp
(C) Perisperm (D) Seed coat

68. Germ pores are actually
(A) Apertures in exine (B) Thick area in intine
(C) Apertures in intine (D) Thin area in intine

69. The plant part which consist of two generations, one within the other, is
(A) Germinated pollen grain (B) Embryo
(C) Unfertilized ovule (D) Seed

70. Which one of the following protects the radicle?
(A) Epicotyl (B) Hypocotyl
(C) Coleoptile (D) Coleorhiza

71. Seed coat of angiosperm seed is also known as
(A) Testa (B) Spore
(C) Hilum (D) Epicotyl

72. Seeds of monocotyledons are
(A) Albuminous or endospermic
(B) Ex-albuminous or non-endospermic
(C) Cotyledon
(D) Perisperm

73. Miniature plants contains one or two seed leaves called
(A) Radicle (B) Plumule
(C) Cotyledon (D) Zygote

74. The seeded fruit of rice and maize is called
(A) Berry (B) Pepo
(C) Siliqua (D) Caryopsis

75. Dormancy caused by hard seed coat is overcome by an
(A) Exposure to abrasion
(B) Exposure to high temperature
(C) Exposure to light
(D) Exposure to cold temperature

76. For which crop, the dormancy is advantageous?
(A) Sesamum (B) Niger
(C) Castor (D) Groundnut

77. Which part of the seed is decomposed by microbes to break dormancy
(A) Hard endocarp and exocarp
(B) Hard seed coat and exocarp
(C) Papery seed coat and exocarp
(D) Epicarp and mesocarp

78. Which one of the following is an appendage?
(A) Hilum (B) Chalaza
(C) Strophiole (D) Funicle

79. When the seed coat forms,the metabolic activity
(A) Slows down (B) Accelerates
(C) Fluctuates (D) Ceases

80. Filiform apparatus is found in
(A) Synergids (B) Anther wall
(C) Secondary nucleus (D) Egg cell

81. Registered seed is directly produced from
(A) Foundation seed (B) Certified seed
(C) Breeder's seed (D) All the above

82. Improved seed distributed to farmers for commercial cultivation is
(A) Nucleus (B) Breeder
(C) Foundation (D) Certified

83. Which of the following is not a commercial class of seed?
(A) Producer (B) Registered
(C) Breeder (D) Foundation

84. International Crop Improvement Association(ICIA) in ____ classified seed into different categories
(A) 1964 (B) 1946
(C) 1963 (D) 1972

85. What is the seed moisture at harvestable maturity?
(A) 10-14% (B) 14-16%
(C) 16-18% (D) 18-20%

86. Choose the correct visual symptom from the following to fix the stage of physiological maturity in rice
(A) Formation of black spot on seed coat
(B) Formation of black spot on endosperm
(C) Formation of black spot on scutellum
(D) Formation of black spot on aleurone layer

87. Harvesting of seed should be done at physiological maturity to avoid harvesting of
(A) Immature and filled seed (B) Mature and ill filled seed
(C) Mature and filled seed (D) Immature and ill filled seed

88. __________ % is seed replacement rate in India
(A) 10-15 (B) 15-20
(C) 20-25 (D) 25-30

89. How SRR is worked out?
(A) (Seed yield / Seed rate) × 100
(B) (Seed yield / Seed rate)
(C) (Qty of actual quality seed sown / Qty of quality seed required) × 100
(D) (Qty of certified seed required / Qty of certified seed sown) × 100

90. When breeder seed policy was introduced?

(A) 1982 (B) 1986

(C) 1984 (D) 1988

91. Which one of the following is not applicable for seed quality control system?

(A) Seed Certification Agency

(B) Seed Law Enforcement Wing

(C) Seed Testing Laboratory

(D) Soil Testing Laboratory

92. Which one of the following is not relevant for seed village with respect to its advantages?

(A) No isolation problem

(B) Easy mechanization

(C) Engage trained labourers

(D) Not easy for inspection by Govt. Agencies

93. Recalcitrant seeds are

(A) Resistant to drying and freezing temperature

(B) Killed by drying and freezing temperature

(C) Both A and B

(D) None of above

94. What is the safe moisture for cereal seed?

(A) 15% (B) 14%

(C) 12% (D) 18%

95. In wheat and rice, the genetic purity of a variety can deteriorate due to

(A) Inbreeding

(B) Mechanical contamination with seeds of other varieties

(C) Delayed segregation

(D) (A) and (B) Both

96. Match the following

(a) Pre-cleaner	1. Removes large materials
(b) Scalper	2. Removes husk
(c) Debeader	3. Removes trash
(d) Huller	4. Removes awn

(A) 3, 1, 4, 2 (B) 4, 2, 3, 1

(C) 2, 3, 4, 1 (D) 2, 3, 4, 1

97. Match the following

(a) Seed fortification	1. Dehydration
(b) Seed priming	2. Thin film coating
(c) Seed polymer coating	3. 10 - 15°C
(d) Seed hardening	4. 20 - 25% mc

(A) 4, 3, 2, 1 (B) 4, 2, 3, 1

(C) 3, 1, 4, 2 (D) 2, 4, 1, 3

98. Match the following

Crop	*Minimum germination (%)*
(a) Jute	1. 65
(b) Bhendi	2. 60
(c) Bengal gram	3. 80
(d) Capsicum	4. 85

(A) 1, 3, 2, 4 (B) 4, 2, 3, 1
(C) 3, 1, 4, 2 (D) 2, 3, 4, 1

99. Match the following

Crop	*Minimum germination (%)*
(a) Wheat	1. 75
(b) Groundnut	2. 85
(c) Watermelon	3. 70
(d) Cotton (hybrid)	4. 60

(A) 4, 3, 2, 1 (B) 2, 3, 4, 1
(C) 1, 3, 4, 2 (D) 2, 4, 1, 3

100. Match the following

Crop	*Working sample (g)*
(a) Cucumber	1. 3
(b) Tomato hybrid	2. 1000
(c) Carrot	3. 15
(d) Broad bean	4. 7

(A) 1, 3, 4, 2 (B) 4, 2, 3, 1
(C) 4, 3, 2, 1 (D) 3, 4, 1, 2

Answer Keys

Q.No.	Ans.	Q.No.	Ans.	Q.No.	Ans.	Q.No.	Ans.	Q.No.	Ans.
1	B	21	C	41	B	61	C	81	A
2	D	22	B	42	D	62	C	82	D
3	B	23	B	43	B	63	C	83	A
4	B	24	C	44	A	64	D	84	B
5	B	25	C	45	D	65	D	85	D
6	B	26	C	46	D	66	D	86	D
7	C	27	D	47	B	67	C	87	D
8	D	28	C	48	A	68	C	88	B
9	D	29	D	49	C	69	C	89	C
10	B	30	B	50	A	70	D	90	B
11	C	31	D	51	C	71	A	91	D
12	B	32	D	52	B	72	A	92	D
13	C	33	A	53	C	73	C	93	C
14	C	34	C	54	D	74	D	94	C
15	C	35	A	55	B	75	A	95	A
16	B	36	C	56	C	76	D	96	A
17	D	37	A	57	C	77	D	97	A
18	C	38	A	58	B	78	C	98	C
19	D	39	C	59	C	79	D	99	B
20	C	40	C	60	A	80	A	100	D

MODEL TEST 3

1. **Which of the following is not true about synthetic seeds?**
   (A) Can be stored for a year without the loss of valuables
   (B) Easy to handle
   (C) Can be directly sown in the soil like natural seeds
   (D) Need hardening in the green house

2. **Recalcitrant seeds are**
   (A) Resistant to drying and freezing temperature
   (B) Killed by drying and freezing temperature
   (C) Both (A) and (B)
   (D) None of above

3. **The encapsulation of somatic embryos can be carried out by**
   (A) Automatic encapsulation process
   (B) Gel complexation
   (C) Both (A) and (B)
   (D) Coating proteins

4. **Seed lot certificate colour is**
   (A) Orange (B) Green
   (C) Purple (D) Orange or green

5. **Seed coat peroxidase test is used for**
   (A) Sorghum (B) Wheat
   (C) Soybean (D) Groundnut

6. **Slow drying seed is**
   (A) Cereals (B) Rapeseed and mustard
   (C) Grasses (D) Pulses

7. **Germination is inhibited by–**
   (A) Red light (B) Blue light
   (C) IR light (D) UV light

8. **What is a characteristic of a monocot?**
   (A) Scattered bundles of vascular tissue in its stem
   (B) Flowers with four or five petals
   (C) Two cotyledons in each seed
   (D) Branching veins in its leaves

9. **A flower's female reproductive parts are called**
   (A) Anthers (B) Filaments
   (C) Pistils (D) Stamen

10. **Which of the following is the male part of the flower?**
    (A) Ovule (B) Pistil
    (C) Petal (D) Stamen

11. Which phrase describes pollination?
(A) The development of mature cones
(B) The development of pollen grains
(C) The joining of sperm and egg cells in an ovule
(D) The transfer of pollen from male to female reproductive structures

12. The process by which a seed sprouts is called
(A) Pollination (B) Dispersal
(C) Fertilization (D) Germination

13. Seeds dispersed by wind are
(A) Heavy (B) Round
(C) Irregular (D) Light

14. These are the male reproductive parts of a flower and are found just within a flower's corolla. Each includes a filament and an anther
(A) Stamen (B) Carpel
(C) Ovary (D) Style

15. These sperm-producing bodies are transferred in order to fertilize a plant
(A) Pollen grains (B) Corolla
(C) Carpel (D) Ovaries

16. This is the female reproductive part of the plant and includes the stigma, the style, and the ovary
(A) Anther (B) Corolla
(C) Stamen (D) Carpel

17. Rupture of seed coat and germination of seed effectively is facilitated by force
(A) Bleeding (B) Grafting
(C) Water pressure (D) Imbibition

18. Female gametophyte develops into an egg after germination with healthy megaspore and produce
(A) Embryo (B) Ovary
(C) Seed (D) Fruit

19. Apomictic embryos in citrus arise from
(A) Maternal sporophytic tissue in ovule
(B) Synergids
(C) Diploid egg
(D) Antipodal cells

20. Fibrous thickenings of hygroscopic nature are found in this part of the anther wall
(A) Middle layers (B) Endothecium
(C) Epidermis (D) Tapetum

21. In a type of apomixis known as adventive embryony, embryos develop directly from the

(A) Zygote
(B) Nucellus or integuments
(C) Synergids or antipodals in an embryo sac
(D) Accessory embryo sacs in the ovule

22. In nature, cleistogamous flowers are

(A) Wind pollinated (B) Insect pollinated
(C) Self pollinated (D) Bird pollinated

23. Which one of the following is surrounded by a callose wall?

(A) Male gamete
(B) Microspore mother cell
(C) Pollen grain
(D) Egg

24. What is the function of germ pore?

(A) Initiation of pollen tube
(B) Absorption of water for seed germination
(C) Emergence of radicle
(D) Release of male gametes

25. Which of the following pairs in angiosperms are diploid and triploid, respectively?

(A) Polar nucleus and secondary nucleus
(B) Microspore mother cell and egg cell
(C) Secondary nucleus and endosperm
(D) Endosperm and antipodal cells

26. The four cells produced in meiosis will have a

(A) 2n number of chromosomes and will differ genetically from each other.
(B) 2n number of chromosomes and will be genetically identical to each other.
(C) n number of chromosomes and will be genetically identical to each other.
(D) n number of chromosomes and will differ genetically from each other.

27. In the F1 generation of a monohybrid cross, the phenotypic ratio would be

(A) 3:1 (B) 1:2:1
(C) 2:1:1 (D) 1:1:2

28. Genetically most pure seed is

(A) Nucleus seed (B) Breeder seed
(C) Certified seed (D) Foundation seed

29. Wind pollinated flowers hang

(A) Upward (B) Downward
(C) Left (D) Right

30. When ovule and nucleus of pollen grain are joined together, we say that ovule is
(A) Pollinated (B) Reproduced
(C) Fertilized (D) Infected

31. In most species of plants, food is stored in
(A) Radicle (B) Plumule
(C) Cotyledon (D) Zygote

32. Egg in carpel is known as
(A) Stigma (B) Style
(C) Ovary (D) Ovule

33. The arrangement of the nuclei in a normal embryo sac in the dicot plants is
(A) 2 + 3 + 3 (B) 3 + 2 + 3
(C) 2 + 4 + 2 (D) 3 + 3 + 2

34. Seed is
(A) Developed ovule
(B) Fertilized and developed ovule
(C) Developed ovary
(D) Fertilized and developed ovary

35. The smallest and lightest seed is that of
(A) Lemna (B) Morus
(C) Orchids (D) Wolffia

36. The largest and heaviest seed is that of
(A) Mango (B) Cocunut
(C) Maldive coconut (D) Rafflesia

37. The seed bearing plants belong to the group of
(A) Pteridophyta (B) Bryophyta
(C) Vascular cryptogams (D) Phanerogams

38. Outer seed coat is
(A) Testa (B) Tegmen
(C) Hilum (D) Funiculus

39. Breeder seed is the progeny of
(A) Foundation seed (B) Registered seed
(C) Nucleus seed (D) Certified seed

40. Certification is not required for
(A) Nucleus seed (B) Breeder seed
(C) Foundation seed (D) Certified seed

41. Cuscuta is an objectional weed of
(A) Cowpea (B) Berseem
(C) Sorghum (D) Lucerne

42. Headquarters of the Union for the Protection of New Varieties of Plants is in
(A) Thailand (B) USA
(C) Denmark (D) Switzerland

43. Seed meant for generation distribution to the farmers for commercial crop production refers to

(A) Foundation seed (B) Breeder seed
(C) Certified seed (D) Nucleus seed

44. The recommended ratio of male rows to female rows is in hybrid pearl millet production

(A) 2 to 4 (B) 1 to 2
(C) 4 to 8 (D) 3 to 6

45. Freedom from inert matter and defective seeds

(A) Genetic purity (B) Physical purity
(C) Defective purity (D) Normal purity

46. Coloured varieties of rice have ________ aleurone layer

(A) Thinner (B) Thicker
(C) Coarse (D) Smooth

47. TDC came into existence in

(A) 1996 (B) 1962
(C) 1926 (D) 1969

48. Contamination permitted in maize is

(A) 1% (B) 2%
(C) 0.10% (D) 0.20%

49. ISTA was established in the year

(A) 1871 (B) 1876
(C) 1921 (D) 1924

50. While sampling from a seed lot of 50 bags, the minimum number of primary sample should be

(A) 5 (B) 10
(C) 15 (D) 20

51. Certified seed of cotton should have minimum germination of

(A) 50% (B) 60%
(C) 70% (D) 80%

52. Types of germination on basis of epicotyl and hypocotyl elongation are

(A) Epigeal germination (B) Hypogeal germination
(C) Epicotyl germination (D) Both (A) and (B)

53. Optimum temperature for seed germination is 25 to

(A) 30 degree celsius (B) 40 degree celsius
(C) 27 degree celsius (A) 35 degree celsius

54. Filiform apparatus is a characteristic feature of

(A) Zygote (B) Suspensor
(C) Egg (D) Synergid

55. A monocarpic plant is one which
(A) Has only one carpel
(B) Flowers and fruits only once in life time and thereafter dies
(C) Produces only one seed
(D) None of these

56. In which one of the following pollination is autogamous?
(A) Cleistogamy (B) Geitonogamy
(C) Xenogamy (D) Chasmogamy

57. Insect pollinated flowers are characterized by
(A) Large number of pollens (B) Dry and smooth pollens
(C) Sticky and rough pollens (D) Heavy pollens

58. 8-nucleate embryo sacs are
(A) Monosporic only (B) Bisporic only
(C) Tetrasporic only (D) Any of these

59. Double fertilization involves
(A) Fertilization of the egg by two male gametes
(B) Fertilization of two eggs in the same embryo sac by two sperms brought by one pollen tube
(C) Fertilization of the egg and the central cell by two sperms brought by different pollen tubes
(D) Fertilization of the egg and the central cell by two sperms brought by the same pollen tube

60. Early blight of potato is ____________ disease
(A) Seedborne (B) Airborne
(C) Soilborne (D) All of these

61. The organo-mercurials fungicides used for dry seed treatment usually contains ________ mercury
(A) 0.5 % (B) 1.0 %
(C) 2.0 % (D) 4.0 %

62. All the following plants can be propagated by leaf cuttings except
(A) Begona (B) Hibiscus
(C) Bryophyllum (D) Sansiveria

63. Which of the following method is commonly practiced for root induction?
(A) Cutting (B) Layering
(C) Grafting (D) Budding

64. Rooting in stem cuttings are stimulated by using
(A) Jasmonic acid (B) ABA
(C) Ethylene (D) IAA

65. Find out the incorrect statement about PPV&FR Act.
i. Enacted by India in 2005 adopting *sui generis* system.
ii. It is in conformity with International Union for the Protection of New Varieties of Plants (UPOV), 1978.

iii. The legislation recognizes the contributions of both commercial plant breeders and farmers in plant breeding activity and also provides to implement TRIPS in a way that supports the specific socio-economic interests of all the stakeholders including private and public sectors and research institutions, as well as resource constrained farmers.

(A) (i) alone incorrect
(B) (i) and (ii) are incorrect
(C) (iii) alone incorrect
(D) All are incorrect

66. Find out the correct statement

i. EDV means a variety which has been essentially derived from genetic engineering
ii. EDV means a variety which has been essentially derived from mutation
iii. EDV means tissue culture derived
iv. EDV means back cross derivative

(A) (i) alone correct (B) (i) and (iii) are correct
(C) (iii) and (iv) are correct (D) All are correct

67. Example of plant in which vegetative propagation is occurred by leaves is called

(A) Cannabis (B) Chrysanthemum
(C) Bryophyllum (D) Brassica

68. A small late planted plant of a biennial root crop is usually dug and stored over winter and replanted the next season for seed production is called

(A) Steckling (B) Planting
(C) Multiplication (D) Stocking

69. Cryopreservation is a technique used for

(A) Crystallization of food
(B) Food packing
(C) Seed saving
(D) Preservation of excess production of vegetables

70. Pick the odd one from the crop and seed rate of pure varieties

	Crop	*Pure varieties (g/ha)*
(A)	Tomato	400–500
(B)	Brinjal	100–200
(C)	Chilli	1000–1250
(D)	Sweet pepper	750–800

71. The capability of a seed to germinate and produce normal seedlings is called

(A) Seed viability (B) Seed longevity
(C) Seed dormancy (D) Compatibility

72. Free flowering and seed setting type of small onion is

(A) CO 1
(B) Co (On) 5
(C) MDU 1
(D) Co (On) 5

73. Which of the following is a seed borne viral disease ?

(A) Rice tungro disease
(B) Bean common mosaic
(C) Black gram common mosaic
(D) Bhendi yellow vein clearing

74. What is the minimum genetic purity standard for hybrid seed production?

(A) 90%
(B) 95%
(C) 99%
(D) 100%

75. Which of the following is used as live barrier in hybrid rice seed production plots?

(A) Sunflower plants
(B) Cotton plants
(C) Daincha plants
(D) Castor plants

76. How many number of border rows to be provided on all sides of hybrid pearl millet seed production plot?

(A) 2
(B) 4
(C) 6
(D) 8

77. Cotton is often cross pollinated crop in which about ______ percent of cross pollination occurs

(A) 25%
(B) 40%
(C) 50%
(D) 75%

78. A flower with essential and non-essential whorls

(A) Incomplete
(B) Irregular
(C) Sessile
(D) Complete

79. In a type of apomixis known as adventives embryony, embryos develop directly from the

(A) Zygote
(B) Nucellus or integuments
(C) Synergids or antipodals in an embryo sac
(D) Accessory embryo sacs in the ovule

80. ______________ are naturally occurring chemical compound containing basic nitrogen atoms

(A) Alkaloids
(B) Tannins
(C) Saponins
(D) Phytins

81. In variety release, the best entries are proposed for which trial?

(A) RYT
(B) PYT
(C) MLT
(D) ART

82. What is the time of anthesis

(A) 9-10 am
(B) 10-11 am
(C) 11 am-12 pm
(D) 8-9 am

83. ________ is the maximum permissible offtypes for 99.8% genetic purity

(A) 0.10% (B) 0.20%

(C) 0.30% (D) 0.50%

84. What is the dosage of GA_3 that is sprayed for enhancing flowering in rice hybrid seed production?

(A) 25 g/ha (B) 50 g/ha

(C) 75 g/ha (D) 100 g/ha

85. Which from the following gametocide is used for cotton?

(A) Zinc methyl arsenate

(B) Maleic hydrazide

(C) FW450

(D) Gibberellin

86. The mechanism of male sterility in self-pollinated crops can be successfully utilised to obtain

(A) Selfed seed

(B) Mutated seed without radiation treatment

(C) Hybrid seed without emasculation

(D) Sterile seed

87. What is the recovery percentage of tomato seeds extracted by fermentation method?

(A) 0.6 - 1.0% (B) 0.8 - 1.2%

(C) 0.9 - 1.2% (D) 1.0 - 1.2%

88. ________ is a suitable season for certified seed production of capsicum?

(A) June – July (B) April – May

(C) June – September (D) February - March

89. The example for cold stratification is

(A) Acacia (B) Albizia

(C) Casuarina (D) Eucalyptus

90. An example for double dormancy is ________.

(A) Acacia (B) Teak

(C) Apple (D) Mango

91. The presence or absence of tannic acid in sorghum seed is determined by ____________ test.

(A) Sodium hydroxide (B) Potassium dichromate

(C) Phenol (D) KOH bleach

92. A clone is a group of individuals obtained through . . .

(A) Self pollination (B) Vegetative propagation

(C) Hybridization (D) Cross pollination

93. Which one of the following is the maximum permitted limit of objectionable weeds in certified seed fields of rapeseed?

(A) 0.01 % (B) 0.02 %

(C) 0.05 % (D) 0.10 %

94. In the simple ovary (of one carpel) there is one common type of placentation known as

(A) Axile (B) Parietal
(C) Central (D) Marginal

95. The number of healthy normal seedlings / number of seed tested × 100 =

(A) Per cent viability (B) Per cent germination
(C) Per cent purity (D) Per cent healthy seed

96. Match the following

List 1	*List 2*
(a) Strawberry	1. Begonia
(b) Offset	2. Runner
(c) Bulbil	3. Pineapple
(d) Crown	4. Gerbera

(A) 4, 2, 1, 3 (B) 2, 3, 1, 4
(C) 3, 1, 2, 4 (D) 4, 3, 2, 1

97. Match the following

List 1	*List 2*
(a) Pomegranate	1. Allium
(b) Recurrent-apomixis	2. Cucumber
(c) Stimulative parthenocarpy	3. Apple
(d) Apogamy	4. Polyembryony

(A) 4, 2, 3, 1 (B) 2, 3, 1, 4
(C) 3, 2, 1, 4 (D) 4, 3, 2, 1

98. Match the following

List 1	*List 2*
(a) Physical dormancy	1. Endogenous dormancy
(b) Physiological dormancy	2. Mechanical dormancy
(c) Bioscarification	3. *Delonix regia*
(d) Hot water treatment	4. *Acacia nilotica*

(A) 4, 2, 1, 3 (B) 2, 3, 1, 4
(C) 2, 1, 4, 3 (D) 4, 2, 3, 1

99. Match the following

List 1	*List 2*
(a) Coriander	1. Cold scarification
(b) Teak	2. Non-deep type
(c) Pines	3. Water leaching
(d) Lettuce	4. Water soaking

(A) 3, 4, 1, 2 (B) 2, 3, 1, 4
(C) 3, 1, 2, 4 (D) 3, 4, 2, 1

100. Match the following

	List 1		*List 2*
(a)	Epigeal germination	1.	Aleurone layer
(b)	Hypogeal germination	2.	Peas
(c)	Monocot	3.	Ash gourd
(d)	Protein	4.	Scutellum

(A) 4, 2, 3, 1 (B) 2, 3, 1, 4
(C) 3, 2, 4, 1 (D) 4, 3, 1, 2

Answer Keys

Q.No.	Ans.	Q.No.	Ans.	Q.No.	Ans.	Q.No.	Ans.	Q.No.	Ans.
1	D	21	B	41	D	61	B	81	D
2	B	22	C	42	D	62	B	82	B
3	C	23	B	43	C	63	B	83	B
4	D	24	A	44	A	64	D	84	C
5	C	25	C	45	B	65	A	85	C
6	D	26	D	46	B	66	A	86	C
7	C	27	A	47	D	67	C	87	A
8	A	28	A	48	A	68	A	88	B
9	C	29	B	49	D	69	C	89	D
10	D	30	C	50	B	70	B	90	B
11	D	31	C	51	B	71	A	91	D
12	D	32	D	52	D	72	D	92	B
13	D	33	B	53	A	73	B	93	D
14	A	34	B	54	D	74	B	94	D
15	A	35	C	55	B	75	C	95	B
16	D	36	C	56	A	76	D	96	B
17	D	37	D	57	C	77	A	97	D
18	A	38	A	58	D	78	D	98	C
19	A	39	C	59	C	79	B	99	A
20	B	40	B	60	C	80	A	100	C

MODEL TEST 4

1. **Root suckers are**
   (A) Insects that suck nutrients from the roots of plants
   (B) Shoots that arise from roots to form new plant
   (C) New plant parts that arise from the branches of fruit trees
   (D) Roots of neighboring plants that share water with the another plant
2. **Cryopreservation is a technique used for**
   (A) Crystallization of food
   (B) Food packing
   (C) Seed saving
   (D) Preservation of excess production of vegetables
3. **In hybridization, this is not part of the technique used**
   (A) Transfer pollen
   (B) Collect pollen
   (C) Emasculate
   (D) Use of two plants to bridge the stems together
4. **The following is not a plant growth regulator**
   (A) Acetic acid (B) Auxins
   (C) Gibberellins (D) Ethylene
5. **Disadvantages of natural vegetative propagation includes**
   (A) Lack of dispersal mechanisms
   (B) No fertilization is involved
   (C) Area is quickly colonized
   (D) Daughter plants can grow quickly
6. **What is the optimum concentration of KNO_3 recommended by ISTA for breaking seed dormancy**
   (A) 0.1% (B) 0.2%
   (C) 0.3% (D) 0.4%
7. **Vivipary means**
   (A) Seed germinates inside the fruit
   (B) Seed germinates as usual still on the parent tree
   (C) Seed dormancy is very long
   (D) Because of immature seeds
8. **Adventitious roots grow from**
   (A) Radicle
   (B) Plumule
   (C) Hypophysis
   (D) Any part of the plant body other than the radicle

9. **Cork cambium is**
(A) Pro meristem (B) Ground meristem
(C) Lateral meristem (D) Intercalary meristem

10. **__________ is an example for hypogeal germination**
(A) Bean (B) Maize
(C) Cowpea (D) Soybean

11. **Which hormone triggers the *denovo* synthesis of hydrolytic enzyme during imbibitions?**
(A) GA (B) ABA
(C) IBA (D) IAA

12. **Which from the following is the progeny of nucleus seed?**
(A) Certified seed (B) Foundation seed
(C) Registered seed (D) Breeder seed

13. **Genetic purity for foundation seed is ________ %**
(A) 100 (B) 99.9
(C) 99.5 (D) 99

14. **Certification is not required for**
(A) Nucleus seed (B) Breeder seed
(C) Foundation seed (D) Certified seed

15. **International Crop Improvement Association (ICIA) in __________ classified seed into different categories**
(A) 1964 (B) 1946
(C) 1963 (D) 1972

16. **A cybrid is**
(A) A sex hybrid
(B) A hybrid formed by cell fusion
(C) A plant produced by cell culture
(D) Symptom to hybrid

17. **A hybrid is always**
(A) Heterozygous (B) Homozygous
(C) Often heterozygous (D) Often homozygous

18. **A maintainer line is**
(A) Male fertile and can restore fertility
(B) Male sterile but can restore fertility
(C) Male sterile but cannot restore fertility
(D) Male fertile but cannot restore

19. **A restorer gene is present in**
(A) Nucleus (B) Cytoplasm
(C) Protoplasm (D) Golgi apparatus

20. **Cross between cytoplasmic male sterile x cytoplasmic male fertile**
(A) Produce sterile F1
(B) Produce fertile F1
(C) Produce fertile, sterile in 1:1 ratio
(D) Produce rectorer F1

21. Hybrid seeds can be produced through open pollination with the use of
(A) Inbreds (B) Clones
(C) Pure lines (D) Male sterile lines

22. Which sterility is exploited in hybrid seed production?
(A) Male genetic sterility
(B) Cytoplasmic sterility
(C) Cytoplasmic genetic male sterility
(D) Genetic sterility

23. Mendok is a gametocide used for inducing male sterility in
(A) Onion (B) Sunflower
(C) Maize (D) Cotton

24. Specific combining ability of the parents should be high in order to develop
(A) Hybrids (B) Composites
(C) Varieties (D) Synthetics

25. To produce hybrid seed we need
(A) A male sterile line, restorer and maintainer
(B) A male fertile line and maintainer
(C) A maintainer and restorer
(D) Only male sterile line

26. At what temperature level, TGSM line becomes fertile
(A) > 34°C / 26°C (day / night) (B) > 32°C / 24°C (day / night)
(C) < 24°C / 18°C (day / night) (D) < 22°C / 16°C (day / night)

27. What is the isolation distance of certified seed production of lucerne?
(A) 400m (B) 300m
(C) 200m (D) 100m

28. Which one of the following substrate is recommended for germination of fodder cowpea?
(A) TP/S (B) BP/TP
(C) TP/BP (D) BP/S

29. The isolation distance for foundation seed production of Cumbu-Napier hybrid
(A) 10m (B) 15m
(C) 20m (D) 25m

30. Headquarters of the Union for the Protection of New Varieties of Plants is in
(A) Thailand (B) USA
(C) Denmark (D) Switzerland

31. At what time, the hand pollination should be carried out to increase seed set in sunflower hybrid seed production?
(A) 5 - 8 am (B) 8 - 11 am
(C) 11am - 1 pm (D) 1 - 4 pm

32. Vernalisation stimulates flowering in

(A) Carrot (B) Turmeric

(C) Elephant foot yam (D) Ginger

33. Ornithophily refers to the pollination by which of the following

(A) Snails (B) Birds

(C) Insects (D) Air

34. Ovule is attached to the placenta of ovary wall by

(A) Funicle (B) Raphae

(C) Hilum (D) Testa

35. Plants with ovaries having only one or a few ovules are generally pollinated by

(A) Birds (B) Butterflies

(C) Bees (D) Wind

36. Sporopollenin is chemically

(A) Protein (B) Fatty substance

(C) Homopolysaccharide (D) Heteropolysaccharide

37. Seed bud is

(A) Plumule (B) Tegmen

(C) Ovule (D) Mesocotyl

38. Tegmen is the name of

(A) Embryo shoot (B) Embryonic node

(C) Cotyledon (D) Inner seed coat

39. Hilum of seed is

(A) Scar of micropyle

(B) Scar of funicule

(C) Area through which pollen tube enters the ovule

(D) Glandular and attractant region of ovule

40. Tigellum is

(A) Embryonic axis (B) Embryonic root

(C) Embryonic shoot (D) Embryonic leaf

41. Seed leaf is

(A) Leaf present over the plumule

(B) Colyledon

(C) Coleoptile

(D) Mesocotyl

42. Through seed germination, a embryo develops into

(A) Complete plant (B) Zygote

(C) Flower (D) Seedling

43. External conditions on which germination of seeds is dependent are

(A) Sufficient food storage and water

(B) Live embryo

(C) Sufficient food storage

(D) Oxygen and water

44. First structure develops from embryo during germination

(A) Shoot (B) Flower
(C) Root (D) Coleoptile

45. Loose smut of wheat is

(A) Internally seed borne (B) Externally seed borne
(C) Soil borne (D) All of these

46. Find out the incorrect statement

Farmer is one who

1. Cultivates crops by cultivating the land himself
2. Cultivates crops by directly supervising the cultivation of land through any other person
3. Breeds the variety

(A) 1 alone (B) 2 alone
(C) 3 only (D) All are incorrect

47. Find out the correct statement

1. Farmers' variety has been traditionally cultivated and evolved by the farmers in their fields
2. Is a wild relative or land race or a variety about which the farmers possess the common knowledge
3. Farmer's variety is exempted from application/ registration fees and his application need not be accompanied with fees, affidavit for terminator technology

(A) 1, 2 and 3 are correct (B) 1 and 2 are correct
(C) 1 alone correct (D) 3 alone correct

48. Seed rate for small onion variety CO (On) 5 is ________.

(A) 8 kg/ha (B) 10 kg/ha
(C) 12 kg/ha (D) 15 kg/ha

49. Seeds of onion are short-lived in storage and is stored at ________ moisture at 5°C

(A) 4% (B) 6%
(C) 8% (D) 10%

50. Physical purity percentage required for carrot seed certification is ________.

(A) 86 % (B) 92 %
(C) 95 % (D) 100 %

51. ________ sugar found in germinating seed in large amount

(A) Maltose (B) Sucrose
(C) Cellobiose (D) Lactose

52. The blotter method of seed health testing detects primarily

(A) Virus (B) Fungi
(C) Bacteria (D) Nematodes

53. Match the crop with designated diseases

	Crop		*Designated diseases*
(1)	Cucurbits	(a)	Early blight, leaf spot
(2)	Potato	(b)	Mosaic
(3)	Bhendi	(c)	Yellow vein mosaic
(4)	Tomato	(d)	Brown rot, root knot nematode

	a	b	c	d
(A)	1	3	2	4
(B)	3	4	2	1
(C)	4	1	3	2
(D)	2	4	1	2

54. Pick the correct statement: As per Seed Act (1966) seed includes

(i) Seed of food crops including edible oil seeds and seeds of fruits and vegetables.

(ii) Seedlings, tubers, bulbs, rhizomes, roots, cuttings, all types of grafts and other vegetatively propagated material for food crops (or) cattle fodder

(A) (i) only (B) (ii) only
(C) Both (i) and (ii) (D) Neither (i) nor (ii)

55. A sound seed certification programme requires

(A) Direct participation of breeder
(B) Use of sophesticated equipment
(C) Classification of seed
(D) Support of law

56. The rudimentary root of the seed or seedling that forms the primary root of the young plant is known as

(A) Rachis (B) Radicle
(C) Rachilla (D) Raceme

57. Example of natural vegetative propagation does not include

(A) Coral (B) Bulb
(C) Corm (D) Rhizom

58. Considering corms of natural vegetative propagation, buds are present at

(A) Front side of corm (B) Backside of corm
(C) Top of the corm (D) End of corm

59. Peas and beans split along the

(A) Pericarp (B) Sutures
(C) Micropyle (D) Radicle

60. For seed dispersal, plants living on river banks depend on

(A) Water (B) Insects
(C) Wind (D) Animals

61. In mango, the
(A) Pericarp is fleshy
(B) Receptacle is fleshy
(C) Placenta is fleshy
(D) Funiculus is fleshy

62. Juicy part of apple is the
(A) Testa
(B) Receptacle
(C) Placenta
(D) Funicle

63. Apple fruit is formed from
(A) A superior ovary
(B) An inferior ovary
(C) Stamen
(D) Petals of flower

64. The flower part removed or made non functional during emasculation is
(A) Anther
(B) Calyx
(C) Corolla
(D) Ovary

65. Consider the statements:
The state seed certification agencies perform the following functions
i. They select seed growers
ii. They carry out the requisite field inspections
iii. They conduct seed test
iv. They guide the seed growers on production, processing and distribution of seeds
(A) i Alone correct
(B) i and ii are correct
(C) i, ii and iii are correct
(D) All are correct

66. Minimun isolation distance for foundation seed production in field crops is
(A) Two metres
(B) Three metres
(C) Four metres
(D) five metres

67. Seed sample taken from laboratory sample for testing is known as
(A) Primary sample
(B) Composite sample
(C) Submitted sample
(D) Working sample

68. Seed health test may be done through
(A) Visual examination
(B) Washing technique
(C) Soaking
(D) Any one of the above

69. Which of the following statements is true for the pollen tube?
(A) It shows only tip growth
(B) It is composed of three non cellular zones
(C) It shows chemostatic movements
(D) It shows radial cytoplasmic streaming

70. A true fruit is the one where
(A) Only ovary of the flower develops into fruit
(B) Ovary and calyx of the flower develops into fruit
(C) Ovary, calyx and thalamus of the flower develops into fruit
(D) All floral whorls of the flower develops into fruit

71. In which of the following crop seed, the globulin is predominant
(A) Sunflower (B) Oats
(C) Sorghum (D) Bread wheat

72. For foundation seed production of ornamental sunflower, the minimum germination prescribed by seed testing laboratories is ______
(A) 75% (B) 65%
(C) 70% (D) 80%

73. Which one of the following is not a fruit?
(A) Tomato (B) Potato
(C) Musk melon (D) Water melon

74. What type of dormancy exist in fig?
(A) Endogenous dormancy (B) Chemical dormancy
(C) Seed coat dormancy (D) Embryo dormancy

75. ________ needs dormancy breaking treatment for germination, as observed during germination test
(A) Fresh seed (B) Imbibed seed
(C) Dead seed (D) Fresh ungerminated seed

76. Which one of the following categories of seeds is the farmer expected to buy afresh every time?
(A) Variety (B) Hybrid
(C) Synthetic (D) Composite

77. What is the seed rate for redgram hybrid seed production?
(A) 20 kg A line and 10 kg B line/ha
(B) 12 kg A line and 5 kg B line/ha
(C) 16 kg A line and 9 kg B line/ha
(D) 4 kg A line and 1 kg B line/ha

78. International Seed Testing Association was organized on
(A) 10-Jul-24 (B) 12-Jul-25
(C) 12-Feb-42 (D) 10-Feb-24

79. ________ is the maximum standard for ODV for foundation seed of sorghum?
(A) None (B) 5/kg
(C) 10/kg (D) 20/kg

80. Which of the following class of seed not under persue of seed certification authority.
(A) Breeder seed (B) Nucleus seed
(C) Foundation seed (D) Certified seed

81. ______ is the minimum moisture content of cotton hybrids for seed certification

(A) 7% (B) 8%
(C) 9% (D) 10%

82. How to prepare 0.5% $CaCl_2$ for soaking of groundnut seeds?

(A) 125 g of $CaCl_2$ in 25 litres of water
(B) 125 g of $CaCl_2$ in 20 litres of water
(C) 125 g of $CaCl_2$ in 15 litres of water
(D) 125 g of $CaCl_2$ in 10 litres of water

83. Which one of the decimal place is followed, if the purity anlaysis is done in working seed sample weighing > 1 g to < 10 g while reporting the results?

(A) 1 (B) 2
(C) 3 (D) 4

84. Which one of the following is not relevant to pure seed fraction?

(A) All botanical varieties of that kind / species
(B) Broken seeds, if the size is > ¾ of the original size
(C) Broken seeds, if the size is > ½ of the original size
(D) Immature, shrivelled seeds

85. 2,3,5 - Triphenyl tetrazolium chloride solution is ______ in colour

(A) Colourless (B) White
(C) Red (D) Pink

86. Which one of the following method is used for mixing and dividing of chaffy seeds?

(A) Spoon method (B) Random cup method
(C) Modified halving method (D) Hand halving

87. Which one of the following light condition is most suitable for non-dormant nature of seed?

(A) Red / far red : high (B) Red / far red : Low
(C) Red / far red : very low (D) Darkness

88. Which one of the following is related to the primary dormancy?

(A) Environmental effects
(B) Genotype effects
(C) Environment and genotype effects
(D) Generation effects

89. The root develops from which part of the germinating seed?

(A) Cotyledons (B) Radicle
(C) Pericarp (D) Plumule

90. Number of field inspections for guinea grass is ______

(A) 1 (B) 2
(C) 3 (D) 4

91. Who has worked out the vigour index?
(A) Hiltner
(B) Moore
(C) Abdul-Baki and Anderson
(D) Mathews and Bradnock

92. Sodium hydroxide test is used to identify red and white varieties of ________
(A) Rice (B) Sorghum
(C) Maize (D) Bajra

93. Which one of the following seed has hilum?
(A) Redgram (B) Castor
(C) Paddy (D) Pearl millet

94. Which one of the following is not matched correctly?
(A) Plumule-dicot only
(B) Two cotyledons-dicot only
(C) One cotyledon-monocot only
(D) Endosperm-monocot and dicot

95. Water makes up about ________ % of a mature seed
(A) One to three (B) Five to twenty
(C) Twenty five to forty (D) Forty five to sixty

96. Match the following

List I	List II
(a) Imbibition	1. Phase III
(b) Protein synthesis	2. Phase I
(c) Cell division	3. Lag phase
(d) Phase II	4. Phase II

(A) 4, 1, 2, 3 (B) 2, 4, 1, 3
(C) 3, 1, 2, 4 (D) 4, 3, 1, 2

97. Match the following

List I	List II
(a) Ribosome	1. ATP production
(b) Mitochondria	2. Photosynthesis
(c) Chlorophyll	3. Water
(d) Vacuole	4. Protein synthesis

(A) 4, 1, 2, 3 (B) 3, 2, 1, 4
(C) 3, 1, 2, 4 (D) 4, 3, 2, 1

98. Match the following

List I	List II
(a) GA	1. Aleurone layer
(b) α Amylase	2. Triggers germination
(c) ABA	3. Stops germination
(d) Thiourea	4. Breaks dormancy

(A) 4, 2, 1, 3 (B) 2, 1, 3, 4
(C) 3, 2, 1, 4 (D) 4, 2, 3, 1

99. Match the following

	List I		List II
(a)	Plant introduction	1.	Sona
(b)	Synthetic variety	2.	Kisan
(c)	Composite variety	3.	Kufri kuber
(d)	Interspecific hybridization	4.	Sonara 64

(A) 4, 1, 2, 3 (B) 2, 3, 4, 1
(C) 3, 1, 4, 2 (D) 4, 2, 3, 1

100. Match the following

	List I		List II
(a)	Mass selection	1.	Resistance breeding
(b)	Heterosis	2.	Asexually propagated crops
(c)	Clonal selection	3.	Cross pollinated crops
(d)	Mutation	4.	Self pollinated crops

(A) 4, 1, 2, 3 (B) 3, 2, 1, 4
(C) 3, 1, 2, 4 (D) 4, 3, 2, 1

Answer Keys

Q.No.	Ans.	Q.No.	Ans.	Q.No.	Ans.	Q.No.	Ans.	Q.No.	Ans.
1	B	21	A	41	A	61	A	81	D
2	C	22	C	42	D	62	B	82	A
3	D	23	A	43	D	63	B	83	C
4	A	24	A	44	C	64	A	84	B
5	A	25	A	45	A	65	D	85	A & B
6	B	26	C	46	C	66	B	86	D
7	B	27	D	47	A	67	D	87	A
8	D	28	D	48	A	68	D	88	C
9	C	29	D	49	B	69	A	89	B
10	B	30	D	50	C	70	A	90	C
11	A	31	B	51	A	71	B	91	C
12	D	32	A	52	B	72	C	92	A
13	C	33	B	53	C	73	B	93	A
14	B	34	A	54	C	74	C	94	A
15	B	35	D	55	A	75	D	95	B
16	B	36	B	56	B	76	B	96	B
17	A	37	C	57	A	77	C	97	A
18	D	38	D	58	C	78	A	98	B
19	A	39	B	59	B	79	C	99	A
20	A	40	A	60	A	80	A	100	D

MODEL TEST 5

1. **Find out the irrelevant statement to the advantages of plant breeders' rights?**
   (A) Breeders have exclusive control over variety
   (B) Breeders get benefit of their variety
   (C) There will be compulsion to purchase fresh seed every year
   (D) PBR leads to improvement in quality because of competition
2. **Onion and garlic are examples of**
   (A) Rhizome (B) Corm
   (C) Stem tuber (D) Bulb
3. **Examples of corm include**
   (A) Gloriosa (B) Canna
   (C) Lallang (D) Ginger
4. **A common example of stem tuber is**
   (A) Ginger (B) Garlic
   (C) Onion (D) Potato
5. **External agencies are not required in**
   (A) Natural vegetative propagation
   (B) Sexual reproduction
   (C) Bisexual reproduction
   (D) Trisexual reproduction
6. **Advantages of artificial methods of vegetative reproduction include**
   (A) Crops with better quality
   (B) Crops with more yield
   (C) Ability to produce plants with desirable qualities
   (D) All of these
7. **In which year, NSC was started ?**
   (A) 1962 (B) 1961
   (C) 1963 (D) 1964
8. **In which crop seed, the aleurone layer is present?**
   (A) Redgram (B) Blackgram
   (C) Greengram (D) Barley
9. **__________ is formed from outer integument**
   (A) Tegmen (B) Testa
   (C) Seed coat (D) Pericarp
10. **Endosperm is**
    (A) 1N (B) 2N
    (C) 3N (D) 4N

11. **In which crop, the caruncle is present?**
(A) Castor (B) Wheat
(C) Redgram (D) Sesamum

12. **In cotton, ____________ is present**
(A) Awn (B) Caruncle
(C) Hairs (D) Aril

13. **Which one of the following has folded cotyledon?**
(A) Safflower (B) Cotton
(C) Sunflower (D) Sesamum

14. **In which one of the dicot seed, a thin layer of endosperm is present?**
(A) Safflower (B) Sunflower
(C) Castor (D) Sesamum

15. **Cucumber, squashes and beans are**
(A) Monocots (B) Dicots
(C) Multicots (D) Ferns

16. **The aleurone layer in maize grain is specially rich in**
(A) Lipids (B) Starch
(C) Proteins (D) Auxins

17. **Cotyledons and testa respectively are edible parts in**
(A) Cashewnut and litchi (B) Groundnut and pomegranate
(C) Walnut and tamarind (D) French bean and coconut

18. **Which one of the following is not matched correctly?**
(A) Plumule-dicot only
(B) Two cotyledons-dicot only
(C) One cotyledon-monocot only
(D) Endosperm-monocot and dicot

19. **A seed develops from a(an)**
(A) Pollen grain (B) Embryo
(C) Ovule (D) Ovary

20. **What would be the number of chromosomes of the aleurone cells of a plant with 42 chromosomes in its root tip cells?**
(A) 21 (B) 42
(C) 63 (D) 84

21. **The larger cells near the micropyle develop into**
(A) Ground meristem (B) Shoot apical meristem
(C) A suspensor (D) Protoderm

22. **Water makes up about ______________ % of a mature seed**
(A) One to three (B) Five to twenty
(C) Twenty five to forty (D) Forty five to sixty

23. **Double fertilization was first discovered by Nawaschin (1898) in**
(A) *Lilium* and *Fritillaria* (B) Mango and sugarcane
(C) Papaya and pea (D) Brassica and candyfruit

24. Hard seeds are commonly found in families of

(A) Amaranthaceae (B) Leguminosae
(C) Alliaceae (D) Solanaceae

25. What is an example for photo dormancy?

(A) *Lactuca* (B) *Phacetia*
(C) *Tectona* (D) *Acacia*

26. Which one of the following is a primary dormancy

(A) Quiescence (B) Endogenous dormancy
(C) Thermodormancy (D) Photodormancy

27. Which one of the following event is a resting stage of seed

(A) Quiescence (B) Endogenous dormancy
(C) Thermodormancy (D) Photodormancy

28. Hypanthodium refers to a special type of

(A) Fruit (B) Flower
(C) Inflorescence (D) Ovary

29. If any somatic cell of sporophyte produces gametophyte without reduction division, it is

(A) Parthenogenesis (B) Apogamy
(C) Apospory (D) Amphimixis

30. In agave, vegetative propagation is carried out through

(A) Bulbils (B) Rhizome
(C) Stolon (D) Sucker

31. In an angiosperm, male plant is diploid and female plant is tetraploid, endosperm will be

(A) Haploid (B) Triploid
(C) Tetraploid (D) Pentaploid

32. A clone is a group of individuals obtained through

(A) Self pollination (B) Vegetative propagation
(C) Hybridization (D) Cross pollination

33. Which one is the reproductive unit having an embryo, reserve food and protective covering

(A) Spore (B) Fruit
(C) Seed (D) Fruitlet

34. In exalbuminous seeds, there is

(A) Nonformation of endosperm
(B) Abundant formation of endosperm
(C) Abundant protein
(D) Consumption of endosperm during development of seed

35. In exalbuminous seed, the food is generally stored in

(A) Endosperm initially but hypocotyls later on
(B) Cotyledons
(C) Endosperm from the beginning
(D) Testa

36. The seed in which reserve food is present in the perisperm
(A) Nymphaea (B) Coconut
(C) Onion (D) Litchi

37. An endospermic seed is
(A) Pea (B) Gram
(C) Castor (D) Bean

38. Seed drying is very important to maintain its
(A) Viability and vigour (B) Protein content
(C) Oil content (D) Chemical composition

39. The equipment used to apply chemicals that involves suspension and wettable powder treatment material in water is
(A) Slury treaters (B) Direct treaters
(C) Panogen treaters (D) Mist-o-matic treaters

40. Fungi and bacteria usually enters through–
(A) Stomata (B) Wounds
(C) Hydathodes (D) Insect puncture

41. Haryali Yojna is related to–
(A) Soil management (B) Water management
(C) Crop management (D) Air management

42. Stripe rust is due to–
(A) *Puccinia Striiformis* f. sp. *tritici*
(B) *P. grantiuis tritici*
(C) *P. graminis recondita*
(D) None

43. Average diameter of bulbet of onion (seed standard) shall not be less than
(A) 2.5 cm (B) 1 cm
(C) 3 cm (D) 2 cm

44. Loose smut of barley is/are–
(A) Internally seed borne (B) Externally seed borne
(C) Both (A) and (B) (D) Not clearly defined

45. The rights which are comes under PPV&FR Act is/are ____________
(A) Breeders' rights (B) Researchers' rights
(C) Farmers' rights (D) All the above

46. Plant breeders rights are operating in
(A) Denmark (B) Netherlands
(C) Germany (D) All of the above

47. Association of Official Seed Certifying Agencies (AOSCA) formerly known as the International Crop Improvement Association in ______ classified seed into different categories
(A) 1964 (B) 1946
(C) 1963 (D) 1972

48. Presently, ICAR has ______________ number of breeder seed production unit

(A) 45 (B) 54
(C) 95 (D) 102

49. The first symbolic importance of seed started after the report of Famine Commission in ________.

(A) 1881 (B) 1890
(C) 1891 (D) 1892

50. Standards of germination for seed certification in chillies is ______.

(A) 70% (B) 90%
(C) 80% (D) 60%

51. Anemochorous seed bearing hair at two ends is

(A) Alstonia (B) Calotropis
(C) Bombax (D) Cotton

52. Hygroscopic hair for fruit dispersal to moist areas occurs in

(A) Sonchus (B) Cotton
(C) *Mucuna gigantea* (D) *Entada gigas*

53. Plumed seeds occur in

(A) *Naravelia* (B) *Taraxacum*
(C) *Calotropis* (D) *Colutea*

54. Parachute mechanism of fruit dispersal is found in

(A) *Bombax* (B) *Taraxacum*
(C) *Holoptelia* (D) *Polygonum*

55. Coconut fruit is dispersed by

(A) Air (B) Animals
(C) Autochory (D) Water

56. In coconut, hydrochory occurs due to presence of

(A) Papery epicarp (B) Fibrous mesocarp
(C) Stony endocarp (D) Liquid endosperm

57. In cutting method of vegetative propagation, cuttings are mainly taken from

(A) Leaves of the parent plant
(B) Roots or stems of parent plant
(C) Shoots of parent plant
(D) Buds of parent plant

58. Artificial methods of vegetative propagation includes

(A) Cloning (B) Grafting
(C) Cuttings (D) Both (B) and (C)

59. Which type of association is found in between entomophilous flower and pollinating agent?

(A) Cooperation (B) Commensalism
(C) Mutualism (D) Coevolution

60. Transfer of pollen grains from the anther to the stigma of another flower of the same plant is called

(A) Geitonogamy (B) Xenogamy
(C) Autogamy (D) Karyogamy

61. When the other floral parts are arranged at the base of the gynoecium, the flower is called

(A) *Perigynous* flower (B) *Epigynous* flower
(C) *Hypogynous* flower (D) None of these

62. Male gametes in angiosperms are formed by the division of

(A) Generative cell (B) Microspore
(C) Microspore mother cell (D) Vegetative cell

63. The aleurone layer in maize grain is specially rich in

(A) Lipids (B) Starch
(C) Proteins (D) Auxins

64. Endosperm is formed during the double fertilization by

(A) Ovum and male gamete
(B) One polar nucleus and one male gamete
(C) Two polar nuclei and one male gamete
(D) Two polar nuclei and two male gametes

65. In angiosperms, pollen tube liberate their male gametes into the

(A) Egg cell (B) Antipodal cells
(C) Central cell (D) Synergids

66. Which is the most useful method of detecting virus infection in seeds?

(A) Seed plating (B) Phago test
(C) ELISA (D) None of these

67. The ideal requirements for good seed are

(A) Genetic and physical purity
(B) Good germination
(C) Freedom from diseases and weeds
(D) All the above

68. The Seeds Act was enacted by the parliament in the year

(A) 1965 (B) 1966
(C) 1967 (D) 1968

69. The test used for quick testing of viability of seeds is

(A) Acid fuchsin test (B) Thiourea test
(C) Tetrazolium test (D) Succinic acid test

70. For breeder seed production, colour of tag used in the field is

(A) Blue (B) Golden yellow
(B) White (D) None of the above

71. Breaking dormancy of hard seed coats seed by mechanical (or) chemical treatments is called as

(A) Stratification (B) Scarification
(C) Sterilization (D) Vernalization

72. Acceleration of ability of flower by a chilling treatment called is called as

(A) Vernalization (B) Devernalization
(C) Sterilization (D) Accumulation

73. ________ are synthetic compounds which reduce the growth of the plants.

(A) Growth stimulant (B) Growth promoters
(C) Growth retardants (D) Growth enhancers

74. The transfer of pollen grains from anther to the stigmatic surface is called as

(A) Fertilization
(B) Pollination
(C) Transformation
(D) Self incompatibility

75. Thermodormancy can be overcome by application of

(A) ABA (B) Cytokinins
(C) Retardants (D) Auxins

76. Pre germination techniques in seeds followed for increasing

(A) Purity (B) Viability
(C) Germination (D) Storability

77. Seedlessness in grapes can be induced by the application of ____ in Anab-e-Shahi.

(A) GA_3 @ 50 ppm (B) GA_3 @ 100 ppm
(C) GA_3 @ 150 ppm (D) GA_3 @ 200 ppm

78. Seed germination in pea is enhanced by _____ treatment

(A) IBA (B) GA
(C) Ethylene (D) MH

79. Cluster been seed contain gum like musilaginous substances called

(A) Galactomannon (B) Gell
(C) Fluid (D) Tanin

80. Under which section of the Indian Seeds Act 1966, the varieties are notified

(A) Sec. 3 (B) Sec. 5
(C) Sec. 7 (D) Sec. 9

81. Seed pelleting is a

(A) Pre-sowing seed treatment
(B) Pre-storage seed treatment
(C) Post-storage seed treatment
(D) Mid-storage seed treatment

82. Which of the following statement is true regarding vegetative propagation?

(A) Vegetative propagation occurs naturally and artificially producing genetically identical plants

(B) Vegetative propagation occurs naturally and artificially producing genetically different plants

(C) Vegetative propagation occurs artificially producing genetically similar plants

(D) Vegetative propagation occurs naturally producing genetically similar plants

83. What is the submitted sample size of cotton varieties

(A) 150 g (B) 160 g

(C) 250 g (D) 350 g

84. For preconditioning of seed for TZ test, slow moistening is given for __________ seeds

(A) Sorghum (B) Soybean

(C) Maize (D) Paddy

85. Which from the following is used for delinting of cotton seeds

(A) 2% KCl (B) 0.5% $CaCl_2$

(C) 2% KH_2PO_4 (D) 100ml H_2SO_4

86. In wheat, production of foundation seed needs an isolation distance of

(A) Three metres (B) Five metres

(C) Ten metres (D) Twenty metres

87. When cereal grains germinate, amylase and other hydrolytic enzymes are secreted by the

(A) Scutellum (B) Suspensor

(C) Aleurone layer (D) Endosperm

88. What method is adopted for moisture determination of seeds which contain volatile oils?

(A) Moisture meter method

(B) Air oven method

(C) Toluene distillation method

(D) Lyophilization method

89. Which one of the following method is used for mixing and dividing of seeds requiring working sample upto 10 g.

(A) Spoon method (B) Random cup method

(C) Modified halving method (D) Hand halving method

90. In which one of the following crops, maximum exploitation of hybrid vigour during the last three decades has been done?

(A) Wheat (B) Rice

(C) Maize (D) Cotton

91. What is the seed recovery percentage in HCl method of extraction of ashgourd?

(A) 4-5% (B) 5-6%
(C) 6-7% (D) 7-8%

92. ____________ counts are taken by the seed certification officer when the area sown for seed purpose is 6-10 acres

(A) 5 counts (B) 6 counts
(C) 7 counts (D) 8 counts

93. Zurich is the headquarters for ____________

(A) AOSA (B) ISST
(C) ISTA (D) UPOV

94. Which one of the following is the optimum maturity stage of pumpkin?

(A) Yellow colouration of fruits
(B) Reddish colouration of fruits
(C) Ashy coating over the fruits
(D) Yellowish browning of fruit rind

95. The isolation distance for foundation seed production of Cumbu-Napier hybrid

(A) 10m (B) 15m
(C) 20m (D) 25m

96. Match the following

List I	List II
(A) Relative heterosis	1. (F1-SH/SH) × 100
(B) Heterobeltiosis	2. (F1-MP/MP) × 100
(C) Standard heterosis	3. (F1-HP/HP) × 100
(D) High percent heterosis	4. (F1-BP/BP) × 100

(A) 4, 2, 1, 3 (B) 2, 4, 1, 3
(C) 1, 3, 2, 4 (D) 4, 2, 3, 1

97. Match the following

List I	List 2
(A) Hand emasculation and pollination	1. Rice
(B) TGMS	2. Maize
(C) Intraspecific hybrids	3. Brinjal
(D) Double top cross	4. Cotton

(A) 4, 2, 1, 3 (B) 2, 3, 4, 1
(C) 3, 1, 4, 2 (D) 4, 3, 1, 2

98. Find out the correct sequence/order of variety release trials

(A) PYT, RYT, CYT, MLT, ART, Variety
(B) CYT, RYT, PYT, MLT, ART, Variety
(C) RYT, PYT, CYT, MLT, ART, Variety
(D) RYT, PYT, CYT, ART, MLT, Variety

99. Which of the following is not relevant to deterioration of varieties?

(A) Mechanical mixtures
(B) Major genetic variation
(C) Natural crossing
(D) Developmental variation

100. Match the following

	List I		List 2
(A)	Seed Control Order	1.	1988
(B)	New Seed Policy	2.	1983
(C)	Consumer Protection Act	3.	2004
(D)	New Seed Bill	4.	1986

(A) 2, 4, 1, 3 (B) 2, 1, 4, 3
(C) 3, 1, 2, 4 (D) 4, 3, 2, 1

Answer Keys

Q.No.	Ans.	Q.No.	Ans.	Q.No.	Ans.	Q.No.	Ans.	Q.No.	Ans.
1	C	21	C	41	B	61	C	81	A
2	D	22	B	42	A	62	A	82	A
3	A	23	A	43	D	63	C	83	D
4	D	24	B	44	A	64	C	84	B
5	A	25	B	45	D	65	D	85	D
6	C	26	B	46	D	66	C	86	A
7	C	27	A	47	B	67	D	87	C
8	D	28	C	48	B	68	B	88	C
9	B	29	C	49	A	69	C	89	B
10	C	30	A	50	A	70	B	90	C
11	A	31	D	51	A	71	B	91	B
12	C	32	B	52	A	72	A	92	B
13	B	33	C	53	C	73	C	93	C
14	C	34	D	54	B	74	B	94	D
15	B	35	B	55	D	75	B	95	A
16	C	36	A	56	B	76	C	96	B
17	B	37	C	57	B	77	B	97	C
18	A	38	A	58	D	78	B	98	C
19	C	39	A	59	C	79	A	99	B
20	C	40	A	60	A	80	B	100	B

MODEL TEST 6

1. **Choose the correct statement**
   1. The Protection of Plant Varieties and Farmers' Rights (PPV&FR) Act enacted by India in 2001.
   2. It is in conformity with International Union for the Protection of New Varieties of Plants (UPOV), 1978.

   (A) 1 & 2 are correct (B) 1 alone correct
   (C) 2 alone correct (D) Both are incorrect

2. **Which of the following factors could lead to variations in the offspring of asexually reproducing organisms?**
   (A) Crossing over (B) Fertilization.
   (C) Mutations (D) Independent assortment

3. **The idea that different pairs of alleles are passed to offspring independently in Mendel's principle of**
   (A) Unit inheritance (B) Segregation
   (C) Independent assortment (D) None of these

4. **The idea that for any particular trait, the pair of alleles of each parent separate and only one allele from each parent passes to an offspring in Mendel's principle of**
   (A) Independent assortment (B) Hybridization
   (C) Segregation (D) None of these

5. **Which is not applicable to vegetatively propagated plants**
   (A) To get true to type plants (B) To protect an elite quality
   (C) To conserve a plant (D) To get a new type of plant

6. **Banana is propagated by**
   (A) Stem cutting (B) Tillers
   (C) Suckers (D) Crown

7. **Aims of plant breeding are to produce**
   (A) Disease-free varieties (B) High-yielding varieties
   (C) Early-maturing varieties (D) All of the above

8. **The desired varieties of economically useful crops are raised by**
   (A) Vernalization (B) Mutation
   (C) Natural selection (D) Hybridisation

9. **Which one of the following is sprayed to pulses for the production of bolder seeds?**
   (A) 0.5% DAP (B) 1% DAP
   (C) 1.5% DAP (D) 2% DAP

10. In cotton, spraying of ______ DAP at boll formation increases seed quality

(A) 0.5% (B) 1%
(C) 1.5% (D) 2%

11. Which from the following contaminant does not cause physical contamination?

(A) Pollen shedders
(B) Weed seeds
(C) Designated diseased plants
(D) Other crop plants

12. Which one of the following crop can be harvested in single harvest?

(A) Cotton (B) Tomato
(C) Rice (D) Chillies

13. Which one of the following is a genetic contaminant in cotton?

(A) Selfed bolls (B) Crossed bolls
(C) Skewed bolls (D) Unopened bolls

14. What is the isolation distance for wheat certified seed production?

(A) 200 (B) 150m
(C) 50m (D) 3m

15. Which from the following is the rabi season?

(A) February – May (B) June – September
(C) October – January (D) November – December

16. For seed production of sunflower, the sowing season is ______ ______

(A) August – September (B) April – August
(C) October – November (D) November – December

17. Which one of the following is the seed rate for bunchy type groundnut?

(A) 10 kg/ha (B) 60 kg/ha
(C) 100 - 130 kg/ha (D) 35-40 kg/ha

18. Which one of the following treatment is adopted for hardening of bajra seeds?

(A) 2% KCl (B) 0.5% $CaCl_2$
(C) 2% KH_2PO_4 (D) 100ml H_2SO_4

19. Which one of the following is the recommended sieve size for size grading of rice seed?

(A) 23/64" (B) 8/64"
(C) 24/64" (D) 27/64"

20. What is the permitted weed seed in redgram?

(A) 5/kg (B) 10/kg
(C) 20/kg (D) 40/kg

21. For _______ seed, 75% germination is permitted as minimum standard

(A) Wheat (B) Cotton
(C) Greengram (D) Castor

22. Production of breeder seed in cotton requires an isolation distance of

(A) 20 metres (B) 30 metres
(C) 50 metres (D) 75 metres

23. Oat is said to be

(A) Cross pollinated crops
(B) Self pollinated crop
(C) Mixed pollinated crop
(D) Often cross pollinated crop

24. Among vegetables, which one of the following crop needs a less isolation distance

(A) Brinjal (B) Crucifers
(C) Tomato (D) Bhendi

25. ____________ is the isolation distance for FS of onion

(A) 50 m (B) 400 m
(C) 1600 m (D) 1000 m

26. Which from the following is the isolation distance for FS of bhendi?

(A) 50 m (B) 400 m
(C) 1600 m (D) 1000 m

27. Which season is a suitable for certified seed production of bittergourd

(A) June – July (B) April – May
(C) June – September (D) February – March

28. Which one of the following is a designated disease for muskmelon?

(A) Phompsis fruit blight (B) Anthracnose
(C) Cucumber mosaic (D) Black leg

29. Which one of the following is the optimum maturity stage of pumpkin?

(A) Yellow colouration of fruits
(B) Reddish colouration of fruits
(C) Ashy coating over the fruits
(D) Yellowish browning of fruit rind

30. What is the function of germ pore?

(A) Initiation of pollen tube
(B) Absorption of water for seed germination
(C) Emergence of radicle
(D) Release of male gametes

31. Which of the following pairs in angiosperms are diploid and triploid, respectively?

(A) Polar nucleus and secondary nucleus
(B) Microspore mother cell and egg cell
(C) Secondary nucleus and endosperm
(D) Endosperm and antipodal cells

32. Which of the following statements about sporopollenin is false?

(A) Exine has apertures called germ pores where sporopollenin is present
(B) Sporopollenin is one of the resistant organic materials
(C) Exine is made up of sporopollenin
(D) Sporopollenin can withstand high temperatures and strong acids

33. Which one of the following is not a correct explanation of cross pollination?

(A) The pollen grains of male flowers are transferred to the stigma of the female flowers
(B) The pollen grains are transferred from one flower to another flower, of another plant of the same species
(C) The pollen grains are transferred from one flower to another flower situated on the same plant
(D) The pollen grains of one flower are transferred to the stigma of the same flower

34. Endosperm of areca/betel nut is

(A) Tetraploid (B) Soft
(C) Smooth (D) Ruminate

35. Integumentary outgrowth present over the micropyle of Castor is

(A) Chalaza (B) Caruncle
(C) Aril (D) Wing

36. Hair of cotton are

(A) Outgrowth of testa
(B) Outgrowth of inner layer of perisperm
(C) Constituents of aril
(D) Formed from caruncle

37. Which is an albuminous seed

(A) Gram (B) Castor
(C) Maize (D) Both (B) and (C)

38. Cotyledon is absent in

(A) *Polygonum* (B) *Cuscuta*
(C) *Santalum* (D) *Shorea*

39. Which are the external conditions required for seed germination

(A) Oxygen, carbon dioxide and suitable temperature
(B) Oxygen, light and suitable temperature
(C) Light, moisture and suitable temperature
(D) Oxygen, moisture and suitable temperature

40. A seed which does not require oxygen for germination is

(A) Pea (B) Rice
(C) Typha (D) Both (B) and (C)

41. A seed which is unable to germinate in the presence of light is

(A) Viscum (B) Onion
(C) Bean (D) Maize

42. The seed in which dormancy can be broken by red light is

(A) Pea (B) Gram
(C) Lettuce (D) Castor

43. Part of the embryo which comes out first during seed germination is

(A) Radicle (B) Plumule
(C) Epicotyl (D) Hypocotyl

44. Which is not essential for seed germination in most cases

(A) Oxygen (B) Light
(C) Suitable temperature (D) Moisture

45. Part of the seed which forms the shoot at the time of germination is

(A) Radicle (B) Cotyledons
(C) Epicotyl (D) Plumule

46. Germination is hypogeal in

(A) Cotton (B) Pea
(C) Castor (D) Bean

47. Vivipary is seed germination

(A) In strong light
(B) Without involving cotyledons
(C) With cotyledons coming above ground
(D) While contained inside the fruit

48. Cotyledons constitute the first pair of leaves in

(A) Castor (B) Maize
(C) Pea (D) Gram

49. Epigeal germination occurs in

(A) Pea (B) Gram
(C) Castor (D) Maize

50. The major storage fungi that affects the food grain is

(A) *Rhizobium* (B) *Mucor*
(C) *Cercospora* (D) *Aspergillus*

51. Find out the correct statement about farmers rights act

1. A farmer who has evolved or developed a new variety is entitled for registration and protection in like manner as a breeder of a variety
2. Farmers variety can also be registered as an extant variety.
3. Farmer shall not be liable to pay any fee in any proceeding before the Authority or Registrar or the Tribunal or the High Court under the Act
4. Farmers are NOT eligible for recognition and rewards for the conservation of Plant Genetic Resources of land races and wild relatives of economic plants

(A) 1, 2, 3 are correct
(B) 1 and 2 are correct
(C) 3 and 4 are correct
(D) All are correct

52. Which of the following are required for germination of seed?

(A) Water
(B) Good quality of soil
(C) Optimum temperature
(D) All of these

53. Global seed vault also referred to as 'Doomsday seed vault' which is located in __________.

(A) U.S.A.
(B) India
(C) Norway
(D) Finland

54. Svalbard Global seed vault is founded by _________.

(A) Cary Fowler
(B) Norman Borlaug
(C) Robert Bakewell
(D) Luther Burbank

55. Headquarters of the Union for the Protection of New Varieties of Plants is in

(A) Thailand
(B) USA
(C) Denmark
(D) Switzerland

56. Improved seed results in:

(i) Better germination
(ii) Vigorous seedling growth
(iii) Higher crop stand
(iv) Better quality of produce
(v) Higher crop yieldd

(A) (i) and (ii) only
(B) (i), (ii), (iii) only
(C) All the above
(D) None of the above

57. What are types of improved seed?

(i) Nucleus seed
(ii) Breeder seed
(iii) Foundation seed
(iv) Registered seed

(A) (i) only
(B) (i), (ii), (iii) only
(C) (iv) only
(D) All the above

58. Match the types of dormancy and its causes

1. Physical	-	I.	Under developed embryo, physiological
2. Morphological	-	II.	Seedcoat impermeable
3. Physiological	-	III.	Under developed embryo
4. Morpho-physiological	-	IV.	Physiological inhibiting mechanism of germination in the embryo

	I	II	III	IV
(A)	2	3	4	1
(B)	1	2	3	4
(C)	2	3	1	4
(D)	3	2	4	1

59. Seed drying is very important to maintain its

(A) Viability and vigour (B) Protein content
(C) Oil content (D) Chemical composition

60. Average diameter of bulbet of onion (seed standard) shall not be less than

(A) 2.5 cm (B) 1 cm
(C) 3 cm (D) 2 cm

61. Germination of seeds in some lettuce varieties is promoted by

(A) Blue light (B) Red light
(C) Green light (D) White light

62. Flow of seeds far away from their parent plants is known as

(A) Dispersion (B) Pollination
(C) Fertilization (D) Reproduction

63. Rubber fruits depend upon seed dispersal mechanism of

(A) Wind (B) Insects
(C) Explosive mechanism (D) water

64. Seed dispersal

(A) Avoids competition (B) Produces mixed population
(C) Promotes cores population (D) All the above

65. Which plant hormone is responsible for saving the crops from falling

(A) Cytokinin (B) Gibberellins
(C) Auxins (D) Ethylene

66. Which of the following hormone is found in gaseous form?

(A) Florigens (B) Abscisic Acid
(C) Ethylene (D) Auxins

67. Conifers cuttings are practiced for propagating pine, spruce etc. They are usually taken during

(A) Early spring (B) Early winter
(C) Late spring (D) Late winter

68. The method of inducing rooting by girdling branches is called air layering. Air layering is also called as

(A) Gootee (B) Chinese layering
(C) Pot layering (D) All of these

69. Which one of the following seed germinates on the plant itself after reaching maturity

(A) Jack (B) Guava
(C) Sapota (D) Pineapple

70. Which one of the following light condition is most suitable for non-dormant nature of seed?

(A) Red / far red : high (B) Red / far red : Low
(C) Red / far red : very low (D) Darkness

71. Which one of the following seeds are bioscarified by sheeps and goats

(A) *Acacia purpea* (B) *Acacia tomentosa*
(C) *Acacia nilotica* (D) *Acacia fistula*

72. Seed priming is done in freshly extracted seeds with ______ __________ to overcome dormancy

(A) Phenyl mercuric acetate (B) Polyethylene glycol
(C) Thiourea (D) GA_3

73. The produce reached a desired cooling before storage is known as

(A) Pre cooling (B) Half cooling
(C) Complete cooling (D) System cooling

74. The plant growth regulator __________ induces transport of soluble nitrogen from intact leaves to localize areas of other leaves

(A) Gibberellic acid (B) Kinetin
(C) Ethylene (D) Abscisic acid

75. _____ is the plant growth regulator used for breaking bud dormancy in potato

(A) IBA (B) GA_3
(C) Ethylene (D) Kinetin

76. Application of ____________ at the tapping panel increases the flow of latex in rubber

(A) Ethrel (B) GA_3
(C) Kinetin (D) ABA

77. The preharvest drop of pome fruits can be reduced by spraying the growth regulator

(A) NAA (B) GA
(C) ABA (D) MH

78. If a parthenocarpic fruit develops even without the stimulus from the pollination, then this phenomenon is referred as

(A) Facultative apomixis (B) Vegetative parthenocarpy

(C) Recurrent apomixis (D) Non recurrent apomixis

79. Development of embryo either from Synergids or antipodal cell is referred to as

(A) Apospory (B) Adventive embryony

(C) Apogamy (D) Diplospory

80. _____________ is a reproductive deficiency of same plants where male organs in hermaphrodite flowers are rendered defunction

(A) Self incompatibility (B) Male sterility

(C) Cleistogamy (D) Chasmogamy

81. Number of plants to be counted per count during field inspection in 'widespaced row crops' seed field

(A) 100 (B) 500

(C) 1000 (D) 1500

82. Which one of the following is a direct vigour test for seeds?

(A) Speed of germination

(B) Tetrazolium test

(C) Accelerated ageing test

(D) Seedling length measurements

83. TZ test is used to estimate

(A) Seed moisture (B) Seed purity

(C) Seed viability (D) Genetic purity

84. Hard seeds are commonly found in families of

(A) Amaranthaceae (B) Leguminosae

(C) Alliaceae (D) Solanaceae

85. As per the ISTA rules, the final count for maize seed germination test has to be taken on

(A) 4th day (B) 7th day

(C) 10th day (D) 14th day

86. Which one of the following crop does not involve pollination at all?

(A) Tomato (B) Bittergourd

(C) Cotton (D) Turmeric

87. Even in absence of pollinating agents, seed setting is assured in

(A) *Salvia* (B) *Fig*

(C) *Commelina* (D) *Zostera*

88. Zooidogamy is seen in

(A) *Cycas* (B) *Gnetum*

(C) *Pinus* (D) Angiosperm

89. Which one of the following is not applicable to the functions of State Seed Certification Agency?
(A) To recommend standards for grow-out test
(B) To certify seeds of any notified varieties
(C) To take sample and inspect seed lots
(D) To grant certificates

90. These are the ovule of a plant that has been fertilized by pollen that can grow into a new plant
(A) Seeds (B) Roots
(C) Stems (D) Leaves

91. Which is the most resistant natural organic material?
(A) Cellulose (B) Pectin
(C) Suberin (D) Sporpopollenin

92. Freedom from inert matter and defective seeds
(A) Genetic purity (B) Physical purity
(C) Defective purity (D) Normal purity

93. Which one of the following is not correct for germinable seed while evaluating TZ test
(A) Plumule unstained
(B) Entire embryo stained bright red
(C) Extremeties of scutellum unstained
(D) Extremeties of radicle tip unstained

94. Which one of the following is not applicable for calculating the seed production requirement while seed planning?
(A) Seed replacement rate
(B) Planting ratio
(C) Seed rate of A, B and R lines
(D) Seed yield of A, B and R lines

95. Which composition of seed absorbs more water faster?
(A) Carbohydrate (B) Fat
(C) Protein (D) Vitamin

96. Match the following

List I	**List 2**
(a) Syngenta	1. Japan
(b) DuPont	2. Germany
(c) Bayer	3. USA
(d) Sakata	4. Switzerland

(A) 4, 2, 1, 3 (B) 2, 3, 1, 4
(C) 3, 1, 2, 4 (D) 4, 3, 2, 1

97. Match the following

	List I	List 2
(a)	PPV & FRA	1. 1963
(b)	NSC	2. 1974
(c)	TDC	3. 2001
(d)	NSP	4. 1969

(A) 4, 2, 3, 1 (B) 2, 1, 3, 4
(C) 3, 1, 4, 2 (D) 3, 4, 2, 1

98. Match the following

	List I	List 2
(a)	Paddy	1. 400 SMR
(b)	Finger millet	2. 150 SMR
(c)	Cotton	3. 8 SMR
(d)	Groundnut	4. 50 SMR

(A) 4, 2, 1, 3 (B) 2, 1, 4, 3
(C) 3, 1, 4, 2 (D) 3, 4, 2, 1

99. Match the following

	List I	List 2
(a)	Finger millet	1. 5 SRR
(b)	Pearl millet	2. 3 SRR
(c)	Sesame	3. 1 SRR
(d)	Groundnut	4. 4 SRR

(A) 4, 2, 3, 1 (B) 2, 1, 4, 3
(C) 1, 3, 2, 4 (D) 4, 3, 2, 1

100. Match the following

	List I	List 2
(a)	Dimpled seed	1. Rice
(b)	Barrier isolation	2. Groundnut
(c)	Blossom end rot	3. Peas
(d)	Blackened plumule	4. Tomato

(A) 4, 2, 3, 1 (B) 1, 2, 3, 4
(C) 3, 1, 4, 2 (D) 3, 4, 2, 1

Answer Keys

Q.No.	Ans.	Q.No.	Ans.	Q.No.	Ans.	Q.No.	Ans.	Q.No.	Ans.
1	A	21	C	41	B	61	B	81	A
2	C	22	C	42	C	62	A	82	C
3	C	23	B	43	A	63	C	83	C
4	C	24	C	44	B	64	D	84	B
5	D	25	D	45	D	65	C	85	B
6	C	26	B	46	B	66	C	86	D
7	D	27	A	47	D	67	B	87	C
8	D	28	C	48	A	68	D	88	A
9	D	29	D	49	C	69	A	89	A
10	D	30	A	50	D	70	A	90	A
11	A	31	C	51	A	71	C	91	D
12	C	32	B	52	D	72	B	92	D
13	A	33	D	53	C	73	C	93	A
14	D	34	D	54	A	74	B	94	A
15	C	35	B	55	D	75	B	95	A
16	B	36	A	56	C	76	A	96	D
17	C	37	D	57	D	77	A	97	C
18	A	38	B	58	A	78	B	98	B
19	A	39	D	59	A	79	C	99	D
20	D	40	D	60	D	80	B	100	C

MODEL TEST 7

1. **Mitosis involves separation of only sister chromatids while meiosis involves?**
   (A) Also separation of only sister chromatids
   (B) Separation of only homologous chromosomes
   (C) Separation of homologous chromosomes as well as sister chromatids
   (D) Separation of sister chromatids twice

2. **The cytoplasm of an animal cell is divided by means of**
   (A) A cleavage furrow
   (B) A cell plate
   (C) A cell membrane formed within the cytoplasm
   (D) Mitosis

3. **The step of mitosis in which chromosomes line up along the equatorial plane of the cell is called**
   (A) Prophase (B) Metaphase
   (C) Anaphase (D) Telophase

4. **An example of a heterozygous, but homogenous population is**
   (A) Pure line (B) Hybrid variety
   (C) Synthetic variety (D) Open pollinated variety

5. **Genetically most pure seed is**
   (A) Nucleus seed (B) Breeder seed
   (C) Certified seed (D) Foundation seed

6. **Multiple effects of a single gene is known as**
   (A) Polyploidy (B) Heteroploidy
   (C) Pleiotropy (D) None of these

7. **Which system is used for hybrid seed production in onion?**
   (A) Dioecious (B) Cytoplasmic male sterility
   (C) Genetic male sterility (D) Self incompatibility

8. **What is seed multiplication ratio in potato and groundnut are respectively?**
   (A) 1:4 and 1:8 (B) 1:8 and 1:4
   (C) 1:8 and 1:10 (D) 1:10 and 1:8

9. **A mating among the following can not lead to heterosis**
   (A) AAbb × aaBB (B) AABBccdd × AABBccdd
   (C) AABBccdd × aabbCCDD (D) aaBBcc × AABBCC

10. Pure line variety is a
(A) Homozygous – Heterogenous
(B) Homozygous – Homogenous
(C) Heterogenous – Heterogenous
(D) Heterogenous – Homozygous

11. Centre of diversity refers to the area where cultivated plant species and/or their wild relatives show
(A) Low competition with unrelated species
(B) Highest variation
(C) High ecosystem diversity
(D) Highest variation and high ecosystem diversity

12. The test cross ratio in complementary gene action is
(A) 3:1 (B) 1:3
(C) 1:2:1 (D) 2:1:1

13. Triticale is
(A) Interspecific cross (B) Intraspecific cross
(C) Intergeneric cross (D) Intrageneric cross

14. A plant breeder wants to develop a disease resistant variety. What should he do first?
(A) Hybridisation (B) Mutation
(C) Selection (D) Production of crop

15. Seed treatment with vitavax is the main controlling method of
(A) Loose smut (B) Rust
(C) Downy mildew (D) All of these

16. The instrument, which is used for sowing of seed with fertilizer together at a time is
(A) Seed drill (B) Dibbler
(C) Seed sowing behind plough (D) Fertilizer cum seed drill

17. Seed treatment is done to control
(A) Soil-borne disease (B) Air-borne disease
(C) Seed-borne disease (D) None of these

18. For variety release, DUS test is must. Which authority is conducting DUS test?
(A) ICAR (B) ICRISAT
(C) PPV & FRA (D) IARI

19. Which one of the following maize variety is a composite variety?
(A) African tall (B) Deccan
(C) Ganga (D) COH1

20. Which one of the following is not relevant with respect to seed quality?
(A) Embryonic plant (B) Non-living organism
(C) A dispersal unit (D) Beginning and end of plant life

21. What is the yield increase, if you use certified seed for sowing

(A) 5-10% (B) 10-15%

(C) 15-20% (D) 20-25%

22. Which one of the following is a designated inseparable other crop seed for wheat?

(A) Oats (B) Pearl millet

(C) Rye (D) Sorghum

23. In wheat and rice, the genetic purity of a variety can deteriorate due to

(A) Inbreeding

(B) Mechanical contamination with seeds of other varieties

(C) Delayed segregation

(D) Both (A) and (B)

24. How SMR is calculated?

(A) Ratio of seed yield and seed rate

(B) Ratio of seed rate and seed yield

(C) Ratio of grain yield and seed rate

(D) Ratio of seed rate and grain yield

25. How SRR is worked out?

(A) (Seed yield / Seed rate) × 100

(B) (Seed yield / Seed rate)

(C) (Qty of actual quality seed sown / Qty of quality seed required) × 100

(D) (Qty of certified seed required / Qty of certified seed sown) × 100

26. What is seed replacement rate for hybrids?

(A) 25% (B) 50%

(C) 75% (D) 100%

27. ________ % is replacement rate in India

(A) 10-15 (B) 15-20

(C) 20-25 (D) 25-30

28. Who from the following has suggested the steps for maintaining genetic purity?

(A) Baskin and Delouche (1972)

(B) Baskin and Baskin (1974)

(C) Hartmann and Kester (1968)

(D) Harrington and Woodstock (1962)

29. Which one of the following is the cause for expression of off types?

(A) Recessive gene expression

(B) Dominant gene expression

(C) Co-dominant gene expression

(D) Cytoplasmic expression

30. What is the exact stage of removal of off types in seed field?
(A) After pollination (B) During fertilization
(C) Before pollination (D) After fertilization

31. Which one of the following is not suitable weather for quality seed production?
(A) Moderate rainfall (B) Moderate wind
(C) Optimum temperature (D) Humid climate

32. For which crop, the isolation distance is more or higher?
(A) Self pollinated crop (B) Self fertilized crop
(C) Cross pollinated crop (D) Often cross pollinated crop

33. What is the symptom of molybdenum deficiency in crucifers?
(A) Black rot (B) Whip tail
(C) Blossom end rot (D) Marsh spot

34. What is the symptom of boron deficiency?
(A) Blossom end rot (B) Hollow heart
(C) Darkened plumule (D) Blotching

35. How many pickings are recommended for gourds?
(A) 3-4 pickings (B) 4-5 pickings
(C) 6-8 pickings (D) 8-9 pickings

36. Which one of the following method is adopted for tomato seed extraction?
(A) Wet seed extraction (B) Dry seed extraction
(C) Acid method (D) Manual extraction

37. What is the pure seed percentage of okra?
(A) 95% (B) 97%
(C) 98% (D) 99%

38. Which one of the following pairs of plant structures has haploid number of chromosomes?
(A) Egg cell and antipodal cells
(B) Megaspore mother cell and antipodal cells
(C) Egg nucleus and secondary nucleus
(D) Nucellus and antipodal cells

39. Which one of the following statements is wrong?
(A) Intine is made up of cellulose and pectin
(B) Vegetative cell is larger than generative cell
(C) Pollen grains in some plants remain viable for months
(D) When pollen is shed at two-celled stage, double fertilization does not take place

40. Consider the following statements and choose the correct option

1. The genetic constitution of a plant is unaffected in vegetative propagation
2. Rhizome in ginger serves as an organ of vegetative reproduction
3. Totipotency of cells enables us to micropropagate plants

(A) Statement 2 alone is true
(B) Statements 1and 2 alone are true
(C) Statements 2 and 3 alone are true
(D) All the three statements 1, 2 and 3 are true

41. Grafting is successful in dicots but not in monocots because the dicots have

(A) Cork cambium
(B) Vessels with elements arranged end to end
(C) Vascular bundles arranged in a ring
(D) Cambium for secondary growth

42. In which one pair, both the plants can be vegetatively propagated by leaf pieces

(A) *Asparagus* and *Bryophyllum*
(B) *Agave* and *Kalanchoe*
(C) *Bryophyllum* and *Kalanchoe*
(D) *Chrysanthemum* and *Agave*

43. Pick out the wrong statement

(A) Double fertilization is unique to gymnosperms and monocotyledons
(B) *Sequoia*, a gymnosperm, is one of the tallest trees
(C) Phaeophyceae members possess chlorophyll a, c, carotenoids and xanthophylls
(D) Moss is gametophyte which consists of two stages namely, protonema stage and leafy stage

44. An organic substance that can withstand environmental extremes and cannot be degraded by any enzyme is

(A) Lignin (B) Cellulose
(C) Cuticle (D) Sporopollenin

45. The coconut water and the edible part of coconut are equivalent to

(A) Mesocarp (B) Embryo
(C) Endosperm (D) Endocarp

47. Infection of smut of maize (*Ustilago maydis*) occurs during

(A) After flowering (B) Before flowering
(C) Vegetative stage (D) Reproductive stage

46. **Choose the incorrect statement**
    1. It is in conformity with International Union for the Protection of New Varieties of Plants (UPOV), 1978.
    2. The legislation do not recognize the contributions of both commercial plant breeders and farmers in plant breeding activity

    (A) 2 is wrong (B) 1 is wrong
    (C) Both are wrong (D) Both are correct

48. **Find out incorrect statement regarding the duration of protection of a registered variety.**
    1. For trees and vines - 18 years
    2. For other crops - 9 years
    3. For extant ie., Notified varieties - 15 years from the date of notification of that variety by the Central Government under section 5 of the Seeds Act, 1966

    (A) 2 alone incorrect. (B) 1 and 2 are incorrect
    (C) 3 alone incorrect (D) All are incorrect

49. **International Seed Testing Association was founded on ________.**
    (A) 1924 (B) 1931
    (C) 1945 (D) 1952

50. **The headquarters of ISTA is located in ________.**
    (A) USA (B) Switzerland
    (C) Denmark (D) Slovenia

51. **Pure Live Seed (PLS) is related to**
    (A) Physical purity (B) Genetic purity
    (C) Germination percentage (D) Contamination

52. **In flowering plants, a second seed coat is known as ________.**
    (A) Integument (B) Aluerone layer
    (C) Tegmen (D) Inner ventral scale

53. **Which of the following is a non endospermic seed ?**
    (A) Cereals (B) Millets
    (C) Orchids (D) Custard apple

54. **The endosperm formed as a part of double fertilization ________.**
    (A) Haploid (B) Diploid
    (C) Triploid (D) Polyploidy

55. **All are dicotyledonous seeds except ________.**
    (A) Mango (B) Castor
    (C) Rice (D) Pea

56. **A gas required for germination of pea seed is ________.**
    (A) Oxygen (B) Nitrogen
    (C) Hydrogen (D) Water vapour

57. Which of the following induces seed dormancy?
(A) ABA (B) Potassium nitrate
(C) Gibberellins (D) Ethylene

58. Protective covering over radicle during seed germination is ____________
(A) Suspensor (B) Coleorhiza
(C) Coleoptile (D) Epithelium

59. For seed samples kept in an incubator for germination test, light is ____________
(A) Always essential (B) Never essential
(C) Not harmful (D) Harmful

60. The two most important factors influencing the lifespan of seeds under storage are
(A) Rainfall and relative humidity
(B) Seed size and moisture content
(C) Climate and temperature
(D) Relative humidity and temperature

61. The litchi seed looses viability after removal from within a period of ____________
(A) Six months (B) Two months
(C) One month (D) One week

62. Tetrazolium test determine the level of activities of enzyme
(A) Catalase (B) Dehydrogenase
(C) Carboxylase (D) Isomerase

63. Murate of potash is
(A) K_2SO_4 (B) KCl
(C) KNO_3 (D) K_2HPO_4

64. All angiosperms
(A) Produce fruits (B) Produce cones
(C) Are tropical (D) Are seedless

65. A fruit dispersed by grazing goats and cows is
(A) Xanthium (B) Cucurbita
(C) *Helianthus annuus* (D) *Brassica campestris*

66. Some plants have seeds with hooks for
(A) Pollination (B) Fertilization
(C) Dispersion (D) Reproduction

67. For seeds, task of fruit is to
(A) Keep them fresh (B) Protect them
(C) Disperse them only (D) Protect and disperse them

68. Seeds in digestive system of animals are
(A) Absorbed in large intestine
(B) Digested later in small intestine
(C) Passed out through anus
(D) Stored in rectum

69. Fruits that are eaten by animals are

(A) Tasteless and dry (B) Juicy and bright colored

(C) Juicy but dull colored (D) Tasteless and dull colored

70. Name the plant hormone which is responsible for the ripening of fruits?

(A) Ethylene (B) Auxin

(C) GA (D) Cytokinin

71. Which plant hormone is helpful in making RNA and protein?

(A) Gibberellins (B) Auxins

(C) Cytokinin (D) Ethylene

72. The best method of control for stored grain pests is

(A) Fumigation (B) Systemic insecticide

(C) Proper drying and storage (D) Biological agents

73. Which of the following statement is true regarding vegetative propagation?

(A) Vegetative propagation occurs naturally and artificially producing genetically identical plants

(B) Vegetative propagation occurs naturally and artificially producing genetically different plants

(C) Vegetative propagation occurs artificially producing genetically similar plants

(D) Vegetative propagation occurs naturally producing genetically similar plants

74. The isolation distance for certified seed production in cole crops is

(A) 1600 m (B) 1000 m

(C) 500 m (D) 100 m

75. Steps in seed multiplication are as follows

(a) i. Breeder seed ii. Foundation seed
iii. Registered seed iv. Certified seed

(b) i. Foundation seed ii. Certified seed
iii. Breeder seed iv. Registered seed

(c) i. Certified seed ii. Breeder seed
iii. Registered seed iv. Foundation seed

(d) i. Breeder seed ii. Foundation seed
iii. Registered seed iv. Certified seed

(A) (a) alone is correct (B) (b) alone is correct

(C) (c) alone is correct (D) (d) alone is correct

76. The seed of rice contains 7-8% protein which is located in

(A) Aleurone layer (B) Husk

(C) Inside the endosperm (D) Embryo

77. Wind pollination is called as
(A) Entomophily (B) Zoophily
(C) Anemophily (D) Hydrophily

78. (A × B) × C is
(A) Single cross (B) Double cross
(C) Three way cross (D) Top cross

79. Cotton is a
(A) Self pollinated crop (B) Cross pollinated crop
(C) Often cross pollinated crop (D) Often self pollinated crop

80. Fertilised matured ovule is
(A) Ovary (B) Fruit
(C) Seed (D) Seedling

81. Genetic tool in hybrid seed production of redgram is
(A) CGMS (B) GMS
(C) Mutation (D) PGMS

82. Chemical treatments with ____ can replace the requirement for low temperature exposure to release from endodormancy
(A) Carbon-di-oxide (B) Hydrogen cyanamide
(C) Hydrogen peroxide (D) Nitrogen oxide

83. Which growth regulator prevents storage sprouting in onion and potato?
(A) Ethylene (B) Maleic hydrazide
(C) Cytokinin (D) Gibberllin

84. Gibberellic acid is most effective gametocide in
(A) Lettuce (B) Okra
(C) Tomato (D) Brinjal

85. Seeds which loose viability when dried to moisture content below critical level, irreversible damages are caused to seed. Such seeds are called
(A) Orthodox seeds (B) Recalcitrant seeds
(C) Quiescent seeds (D) Viviparous seeds

86. Seed disinfection is done with
(A) 0.01% mercuric chloride (B) 2% calcium chloride
(C) 1% ethonal (D) 10% methyl alcohol

87. Flowering responses of plants to temperature is known as
(A) Temperature co-efficient (B) Thermodynamics
(C) Vernalisation (D) Thermocelic

88. Albuminous seeds are characterized by
(A) Having endosperm but absence of thick cotyledons
(B) Having thick cotyledons but absence of endosperm
(C) Having thick endosperm and thick cotyledons
(D) Having thick endosperm and absence of cotyledons

89. The secretion of _______, causes rapid closure of stomata

(A) GA (B) ABA
(C) IAA (D) IBA

90. ______________ is not an ethylene inhibitor

(A) Aminoethoxy vinyl-glycine (AVG)
(B) Thidiazuton
(C) Silver thiosulfate
(D) Methylcyclopropene

91. Presence of which of the following in a seed is associated with dormancy

(A) Starch (B) GA
(C) Ethylene (D) ABA

92. The maximum temperature for safe drying of orthodox seeds is

(A) 40°C (B) 41°C
(C) 42°C (D) 43°C

93. Which one of the following is seed upgrading machine?

(A) Debearder (B) Destoner
(C) Pre-conditioning/scalper (D) Gravity separator

94. The priming agent, which does not penetrate the cell wall is

(A) PEG (B) Mannitol
(C) Mannitol (D) Halogens

95. How many plants are to be observed for one count in the case of cotton seed crop?

(A) 50 plants (B) 100 plants
(C) 150 plants (D) 200 plants

96. Match the following

List I	**List II**
(a) GA spray	1. Green gram
(b) DAP spray	2. Rice
(c) Withholding water	3. Sunflower
(d) Rubbing	4. Sorghum

(A) 4, 2, 1, 3 (B) 2, 1, 4, 3
(C) 1, 3, 2, 4 (D) 4, 2, 3, 1

97. Match the following

List I	**List II**
(a) Shedding tassel	1. Tomato
(b) Pollen shedders	2. Rice
(c) Rope pulling	3. Sorghum
(d) Selfed fruit	4. Maize

(A) 4, 2, 1, 3 (B) 2, 4, 1, 3
(C) 4, 3, 2, 1 (D) 3, 4, 2, 1

98. Match the following

List I	List II
(a) Nitrogen	1. Amino acids
(b) Potassium	2. Disease resistance
(c) Magnesium	3. Lodging
(d) Sulphur	4. Chlorophyll

(A) 3, 2, 4, 1 (B) 2, 1, 3, 4
(C) 3, 2, 1, 4 (D) 4, 3, 1, 2

99. Match the following

List I	List 2
(a) Boron	1. Chlorosis
(b) Phosphorus	2. Pollen fertility
(c) Calcium	3. Red purple leaf
(d) Iron	4. Dieback

(A) 2, 4, 1, 3 (B) 2, 1, 3, 4
(C) 3, 1, 2, 4 (D) 2, 3, 4, 1

100. Match the following

List I	List II
(a) Bulb to seed	1. Bhendi
(b) Seed to seed	2. Onion
(c) Stump method	3. Cauliflower
(d) Head to seed	4. Cabbage

(A) 4, 2, 3, 1 (B) 2, 1, 4, 3
(C) 1, 3, 2, 4 (D) 4, 3, 1, 2

Answer Keys

Q.No.	Ans.	Q.No.	Ans.	Q.No.	Ans.	Q.No.	Ans.	Q.No.	Ans.
1	C	21	C	41	D	61	D	81	B
2	A	22	A	42	C	62	B	82	B
3	B	23	A	43	A	63	B	83	B
4	B	24	A	44	D	64	A	84	A
5	A	25	C	45	C	65	A	85	B
6	C	26	D	46	A	66	C	86	A
7	B	27	B	47	C	67	D	87	C
8	A	28	C	48	A	68	C	88	A
9	B	29	A	49	A	69	B	89	B
10	B	30	C	50	B	70	A	90	B
11	D	31	D	51	C	71	C	91	D
12	B	32	C	52	C	72	A	92	D
13	C	33	B	53	C	73	A	93	D
14	C	34	C	54	C	74	A	94	A
15	D	35	B	55	C	75	A	95	B
16	D	36	C	56	A	76	A	96	B
17	C	37	D	57	A	77	C	97	C
18	C	38	A	58	B	78	C	98	A
19	A	39	D	59	C	79	C	99	D
20	B	40	D	60	D	80	C	100	B

MODEL TEST 8

1. **Find out the irrelevant statement about the disadvantages of plant breeders' rights**
   (A) It will promote monopoly
   (B) It will encourage unhealthy practices
   (C) It may lead to increase in prices
   (D) PBR helps in enrichment of genetic resources

2. **An allele is**
   (A) Another word for a gene
   (B) A homozygous genotype
   (C) A heterozygous genotype
   (D) One of several possible forms of a gene

3. **Phenotype refers to the ________________ of an individual**
   (A) Genetic makeup (B) Actual physical appearance
   (C) Recessive alleles (D) None of these

4. **The first generation progeny of a hybrid are called**
   (A) Pure lines (B) Segregates
   (C) Parents (D) F1 plants

5. **The production of double cross hybrid in maize was first suggested by**
   (A) Shull (B) East
   (C) Jones (D) Hull

6. **Who has listed out the factors of apparent and real deterioration of varieties**
   (A) Kadam (1942) (B) Bateman (1947)
   (C) Dettann (1953) (D) Martin (1942)

7. **Which one of the following crop is a self pollinated crop?**
   (A) Tomato (B) Bitter gourd
   (C) Cotton (D) Turmeric

8. **In which one of the family, the protogyny mechanism exist?**
   (A) Malvaceae (B) Graminaceae
   (C) Magnoliaceae (D) Leguminosae

9. **__________ is an example for homogeny**
   (A) Maize (B) Pearl millet
   (C) Mirabilis (D) Sapota

10. **__________ avoids cross pollination**
    (A) Hidden style and ovary (B) Hidden filament and pedicel
    (C) Hidden stamen and stigma (D) Hidden corolla and calyx

11. __________ is an example for pollination by beetles
(A) *Erythrina* (B) *Bombax*
(C) Mango (D) Coconut

12. An example for trimonoecious crop is
(A) Brinjal (B) Banana
(C) Sunflower (D) Ailanthus

13. At the end of pollen tube there are two nuclei called
(A) Sperm and ovum
(B) Tube nucleus and generative nucleus
(C) Tube nucleus and sperm
(D) Generative nucleus and stigma

14. Flowers like night queen, jasmine emit a sweet scent to attract insects like
(A) Honey bees (B) Butterflies
(C) Beetles and moths (D) All the above

15. Genetic self-incompatibility tends to increase
(A) Pollen dispersal (B) Inbreeding
(C) Pollination (D) Cross pollination

16. The female gametophyte of an angiosperm is the
(A) Embryo sac (B) Ovule
(C) Megaspore mother cell (D) Egg Cell

17. If a flower is to be pollinated by birds, it should be
(A) Blue (B) White
(C) Yellow (D) Orange

18. Which of the following is not characteristic of primitive flowers?
(A) Flowering parts are numerous
(B) Flowers are arranged in definite whorls
(C) Petals and sepals are similar
(D) Unfused flower parts

19. When one plants potatoes, the "seed pieces" are parts of
(A) Stolons (B) Adventitious
(C) Rhizomes (D) Suckers

20. Pollen grains on the stigma of a flower reach the egg by forming a
(A) Cotyledon (B) Seed
(C) Pollen tube (D) Carpel

21. In custard apple
(A) Self pollination is likely high
(B) Flowers are dioecious
(C) Anthers and stamens may mature at different times
(D) Flowers are monoecious

22. According to position of the ovule within the ovary an ovule may be

(A) Ascending (B) Descending
(C) Horizontal (D) Vertical

23. An embryo sac of angiosperms generally contains

(A) Two nuclei (B) Three nuclei
(C) Four nuclei (D) Eight nuclei

24. Epicalyx is the characteristic of the family

(A) Gramineae (B) Solanaceae
(C) Malvaceae (D) Compositae

25. Flower is specialized

(A) Shoot (B) Leaf
(C) Stem (D) Tendril

26. Generally embryosac is

(A) Monosporic (B) Bisporic
(C) Tetrasporic (D) Octosporic

27. In cruciferae, the stamens are said to be

(A) Didynamous (B) Tridynamous
(C) Tetradidynamous (D) Decadynamous

28. In potato, the inflorescence is

(A) Helicoid (B) Scropioid cyme
(C) Umbel (D) Raceme

29. Megaspore mother cell produces four megaspores out of which usually

(A) Only one megaspore is functional and the remaining three megaspores abort
(B) Two megaspores are functional and two megaspores abort
(C) Three megaspores are functional and fourth megaspore abort
(D) All the four megaspores are functional

30. The mode of arrangement of ovules in the cavity of ovary is known as

(A) Pollination (B) Aestivation
(C) Placentation (D) Fertilization

31. Umbel inflorescence is characteristic of the family

(A) Mustard family (B) Okra family
(C) Pea family (D) Coriander family

32. When the ovule bends along the funicle that the micropyle lays close to the hilum, the ovule is said to be

(A) Orthotropous (B) Anatropous
(C) Amphilotropous (D) Campylotropous

33. What do you name, if the ovules are arranged at the base of the ovary?
(A) Axile placentation (B) Marginal placentation
(C) Basal placentation (D) Parietal placentation

34. A flower is having a reproductive structure that is either functionally male or functionally female. Which one of the following is correct for this statement?
(A) Hermaphrodite flower (B) Imperfect flower
(C) Staminate flower (D) Pistillate flower

35. The number of outer layers present in each ovule are
(A) 2 (B) 3
(C) 4 (D) 5

36. Sporopollenin is chemically
(A) Protein (B) Fatty substance
(C) Homopolysaccharide (D) Heteropolysaccharide

37. Cauliflory is flowers produced
(A) An old dormant bud (B) On young bud
(C) On the axil (D) On the branch

38. Each stamen is differentiated into
(A) Style (B) Stigma
(C) Ovule (D) Anther

39. The pollen along with pollen tube is known as
(A) Microspore (B) Pollen grain
(C) Megaspore (D) Male gametophyte

40. The process of megasporogenesis occurs in
(A) Embryosac (B) Egg cell
(C) Pollensac (D) Ovule

41. The first cell division of the plant zygote is typically
(A) Longitudinal (B) Diagonal
(C) Transverse (D) Radial

42. Nucellar polyembryony is reported in species of
(A) Brassica (B) Citrus
(C) Gossypium (D) Triticum

43. Development of embryo either from synergids or antipodal cell is referred to as
(A) Apospory (B) Adventive embryony
(C) Apogamy (D) Diplospory

44. Where MLT is conducted?
(A) KVK's of SAU's
(B) Research stations of SAU's
(C) Central Agricultural Universities
(D) ICAR Institutes

45. Application of potash increases

(A) Resistance to water logging
(B) Frost resistance in plants
(C) Disease resistance in plants
(D) None of these

46. The functions of PPV&FR is

1. Registration of new plant varieties, essentially derived varieties (EDV), extant varieties
2. Developing DUS (Distinctiveness, Uniformity and Stability) test guidelines for new plant species
3. Maintenance of National Gene Bank
4. Issuing of seed certification for foundation and certified seed

(A) 1,2 and 3 are correct (B) 2 and 3 are correct
(C) 1 and 4 are correct (D) All are correct

47. __________ is a method of handling dormant seeds in which imbibed seeds are subjected to a period of chilling or warm temperatures to alleviate dormancy conditions in the embryo

(A) Stratification (B) Scarification
(C) Impaction (D) Ignition

48. Arrange the processes going on in seed during germination in correct order

(1) Absorption of water
(2) Hydrolysis of stored food into soluble form
(3) Secretion of enzymes and hormones
(4) Translocation of soluble foods and hormones to the growing points

(A) 1 – 2 – 3 – 4 (B) 4 – 3 – 2 – 1
(C) 1 – 3 – 2 – 4 (D) 1 – 4 – 2 – 3

49. Identify the odd one

(i) Leaf vein cutting : *Begonia*
(ii) Leaf edge cutting : *Pepromia*
(iii) Soft wood cutting : *Coleus*
(iv) Root cutting : *Guava*

(A) (i) only (B) (ii) and (iii) only
(C) (i), (ii) and (iv) only (D) None of these

50. Apple seed dormancy can be released by low temperature treatment by storing the seeds at ________

(A) 5°C (B) 10°C
(C) 15°C (D) 20°C

51. In seeds of *Calluna vulgaris*, dormancy is broken by

(A) Fire (B) Water
(C) Acid (D) Light

52. Endogenously controlled, but environmentally imposed and it is the temporarily suspension of growth accompanied by reduced metabolic activity and relatively independent of ambient environmental conditions

(A) Simpson (B) Amen
(C) Harper (D) Harrington

53. Pick the correct statement

(i) Nuclear - where the endosperm nucleus undergoes several divisions prior to cell wall formation
(ii) Cellular - in which there is no free nuclear phase
(iii) Helobial - where the free nuclear division is preceded, and is followed by cellularization

(A) (i) only (B) (i) and (ii) only
(C) (iii) only (D) All the above

54. The fruits of tribulus are dispersed by

(A) Water (B) Wind
(C) Animals (D) Explosive mechanism

55. Seeds of cinchona are dispersed by

(A) Wind because they are very small and light
(B) Wind because they are winged
(C) Water because they possess aerenchyma
(D) Animals as they have barbs

56. Hooks and barbs are helpful in dispersal of

(A) Viscum (B) Boerhaevia
(C) Plumbago (D) Bidens

57. Wide distribution of coconut in coastal areas of different countries is due to

(A) Its fruits can float on water
(B) It is economically very important
(C) It requires saline soil
(D) It yields oil

58. Fruit of xanthium is zoochorous as it has

(A) Sticky surface (B) Hooks
(C) Edible pulp (D) Sticky pulp

59. Cotton and madar (calotropis) seeds are dispersed by air as they have

(A) Pappus (B) Wings
(C) Hair (D) Flattened

60. Anemochorous seeds and fruits are

(A) Round in shape (B) Flattened
(C) Light in weight (D) Without any bright colour

61. Orchid seeds are usually

(A) Small, light and dry (B) Small and sticky
(C) Large and sticky (D) Large and heavy

62. Which would have wider distribution, those distributed by
(A) Spores (B) Vegetative means
(C) Seeds (D) Fruits

63. Legumes depend upon seed dispersal mechanism of
(A) Wind (B) Insects
(C) Explosive mechanism (D) Water

64. Seeds of orchis/orchids are
(A) Large and heavy (B) Small and hairy
(C) Small and light (D) Large and sticky

65. Seeds dispersed by animals and passing through their alimentary canal are of plant
(A) Cordia (B) Viscum
(C) Loranthus (D) Peepal

66. Fruits having sticky pulp for zoochorous dispersal are of
(A) Guava (B) Banyan
(C) Cordia (D) *Cleome viscosa*

67. A tuft of hair present at one or both ends of the seed (for dispersal) is called
(A) Propeller (B) Coma
(C) Pappus (D) Wing

68. Which one bears coma for anemochory
(A) *Cinchona* (B) *Calotropis*
(C) *Alstonia* (D) Both (B) and (C)

69. Pappus for parachute mechanism of anemochory develops from
(A) Calyx (B) Fruit wall
(C) Seed wall (D) Style

70. Seeds of *Ruellia tuberosa* are disseminated by
(A) Censer mechanism (B) Parachute mechanism
(C) Jaculator mechanism (D) Explosive mechanism

71. Which is total root parasite?
(A) Orobanche (B) Conophalis
(C) Epitagus (D) All

72. Pure seed fraction in *Abelmoschus esculentus* is
(A) 98% (B) 95% (C) 99% (D) 96%

73. The production of high quality and uniform embryos has been limited to only
(A) Carrot (B) Alfalfa
(C) Both (A) and (B) (D) Sandalwood

74. Encapsulation is necessary to produce and protect synthetic seeds. The encapsulation is carried out by various types of hydrogels, which are
(A) Soluble in water (B) Soluble in organic solvents
(C) Insoluble in water (D) Insoluble in organic solvents

75. A flower is brightly coloured, scented and secrete nectar. It is most probably

(A) Pollinated by insects (B) Sterile
(C) Pollinated by wind (D) An insectivorous plant

76. If any somatic cell of sporophyte produces gametophyte without reduction division, it is

(A) Parthenogenesis (B) Apogamy
(C) Apospory (D) Amphimixis

77. Where the food reserves are stored in wheat?

(A) Megagametophyte (B) Embryo
(C) Cotyledons (D) Perisperm

78. Which one of the following is not relevant for deterioration of varieties due to natural crossing

(A) Natural crossing with desirable plants
(B) Natural crossing with undesirable plants
(C) Natural crossing with diseased plants
(D) Natural crossing with off-type plants

79. In self-incompatibility system

(A) Pollen not produced
(B) Self pollination takes place
(C) Pollen does not germinate on self pollination
(D) Cross pollination takes place

80. __________ stage is ecommended for harvesting of marigold for seed purpose

(A) Dried heads (B) Dried pods
(C) Dark purple (D) Orange, red stage

81. When only one fruit develops from the single ovary of a flower it is said to be a

(A) Simple fruit (B) Aggregate fruit
(C) Multiple fruit (D) Composite fruit

82. Seed dormancy is commonly existing in __________ seeds

(A) Avocado (B) Apple
(C) Litchi (D) West

83. Which one of the following is not relevant for seedling evaluation?

(A) Normal seedlings (B) Abnormal seedlings
(C) Hard seeds (D) Fresh seeds

84. Grow out test must be conducted for ______________.

(A) Green gram (B) Finger millet
(C) Cotton (D) Horse gram

85. Planting ratio for cotton hybrid seed production is

(A) 5:2 (B) 8:2
(C) 5:1 (D) 2:1

86. Which from the following gametocide is used for rice?
(A) Ethrel (B) Maleic hydrazide
(C) FW450 (D) Gibberellin

87. ISTA publishes __________
(A) Journal of Seed Technology (B) Seed Science and Technology
(C) Seed Research (D) Journal of Seed Research

88. Which of the following seed crops require a minimum gap of two seasons while selecting the land for seed production?
(A) Groundnut (B) Berseem
(C) Rice (D) Sesamum

89. Progeny of a hybrid plant
(A) Breds true (B) Resembles maternal plant
(C) Resembles paternal plant (D) Segregation

90. What is the maximum permitted limit of objectionable weeds at field level in the case of foundation seed production of rice?
(A) 0.01% (B) 0.02%
(C) 0.05% (D) 0.1%

91. Which one of the following point not suitable to the duties of seed analysts?
(A) To test the seed sample received from seed certification officer
(B) To send the test results to the person concerned
(C) To submit report to state government from time to time
(D) To take seed samples of any notified variety from sale point

92. _______ is the minimum standard for germination of certified seed of groundnut
(A) 60% (B) 65%
(C) 70% (D) 75%

93. The validity period for certified seed is __________
(A) 6 months (B) 9 months
(C) 12 months (D) 15 months

94. ____________ is a designated disease for brinjal
(A) Phompsis fruit blight (B) Anthracnose
(C) Cucumber mosaic (D) Black leg

95. Seed soaking in ___________ is called as biopriming
(A) Water (B) KCl
(C) *Pseudomonas* (D) PEG

96. Which crop is 'queen of cereals'?
(A) Rice (B) Maize
(C) Sorghum (D) Pearl millet

97. How many pollen bearing anthers are present in maize ?
(A) 5000 (B) 4000
(C) 6000 (D) 3000

98. What is the optimum temperature for potato growth and development?

(A) 5 and 10°C (B) 10 and 15°C
(C) 15 and 20°C (D) 15 and 25°C

99. What is the optimum temperature for seed germination in true potato seed production?

(A) 5 to 20°C (B) 10 to 20°C
(C) 15 to 20°C (D) 20 to 25°C

100. Match the following

Flower crops	***Minimum germination (%)***
(a) Petunia	1. 60
(b) Marigold	2. 50
(c) Chrysanthemum	3. 70
(d) Aster	4. 75

(A) 4, 3, 2, 1 (B) 4, 2, 3, 1
(C) 1, 3, 4, 2 (D) 2, 4, 3, 1

Answer Keys

Q.No.	Ans.	Q.No.	Ans.	Q.No.	Ans.	Q.No.	Ans.	Q.No.	Ans.
1	D	21	C	41	C	61	A	81	A
2	D	22	A	42	B	62	D	82	B
3	B	23	D	43	C	63	C	83	D
4	B	24	C	44	B	64	C	84	C
5	C	25	A	45	C	65	D	85	B
6	A	26	B	46	A	66	C	86	A
7	A	27	C	47	A	67	B	87	B
8	B	28	A	48	C	68	D	88	A & B
9	C	29	A	49	D	69	A	89	D
10	C	30	C	50	A	70	C	90	A
11	C	31	D	51	D	71	D	91	D
12	B	32	B	52	B	72	C	92	C
13	C	33	C	53	D	73	C	93	B
14	D	34	B	54	C	74	A	94	A
15	D	35	A	55	B	75	D	95	C
16	C	36	B	56	D	76	C	96	B
17	D	37	D	57	A	77	D	97	C
18	B	38	D	58	B	78	A	98	D
19	C	39	D	59	C	79	C	99	A
20	C	40	D	60	C	80	A	100	A

MODEL TEST 9

1. **Which one of the following is an example for two line hybrid seed production?**
   (A) Sesamum (B) Jute
   (C) Groundnut (D) Cotton
2. **An example for chemical hybridizing agent is**
   (A) PEG (B) Mannitol
   (C) MH (D) Calcium chloride
3. **Which from the following is interspecific hybrid?**
   (A) Varalakshmi (B) Suguna
   (C) Deccan (D) Ganga
4. **__________ hybrid maize is an example for double cross hybrid**
   (A) Jawahar (B) Deccan
   (C) COH1 (D) Ganga
5. **Which from the following crop is an example for GMS method of hybrid seed production?**
   (A) Rice (B) Redgram
   (C) Maize (D) Cotton
6. **Which method is employed for bittergourd hybrid seed production?**
   (A) Pinching of staminate flowers + hand pollination
   (B) Hand emasculation + manual pollination
   (C) GMS + bee pollination
   (D) CMS + natural pollination
7. **In ________ hybrid seed production, the barrier isolation is used**
   (A) Sorghum (B) Pearl millet
   (C) Rice (D) Redgram
8. **Which from the following is the seed rate for bhendi hybrid seed production?**
   (A) 20 kg A line & 10 kg B line/ha
   (B) 12 kg A line & 5 kg B line/ha
   (C) 16 kg A line & 9 kg B line/ha
   (D) 4 kg A line & 1 kg B line/ha
9. **__________ % urea solution can be sprayed to late parent of sorghum hybrid seed production for enhancing flowering**
   (A) 1 (B) 2
   (C) 3 (D) 4

10. Which system is used for hybrid seed production in onion?
(A) Dioecious (B) Cytoplasmic male sterility
(C) Genetic male sterility (D) Self incompatibility

11. The hybrids developed by Government Agencies or Government Institutions and Agricultural Universities are called
(A) Private hybrid (B) Institutional hybrids
(C) Public hybrids (D) Government

12. ____________ deteriorates drastically upon inbreeding
(A) Carrot (B) Okra
(C) Brinjal (D) Broccoli

13. 2n denotes :
(A) Number of homozygote produced by selfing a F1
(B) Number of types of gametes produced by F1
(C) Number of types of heterozygote produced
(D) Number of types of male gametes produced by F1

14. Indian cotton is
(A) *G. herbaceum* (B) *G. arboreum*
(C) *G. hirsutum* (D) *G. barbadense*

15. Sugar turn out from cane in India
(A) Leachate test (B) GADA test
(C) ATPase test (D) Exhaustion test

16. Which of the following vigour test method estimates decarboxylation activities?
(A) Leachate test (B) GADA test
(C) ATPase (D) Exhaustion test

17. For seed samples kept in an incubator for germination test, light is
(A) Always essential (B) Never essential
(C) Not harmful (D) Harmful

18. Seedlessness in fruits is called as
(A) Parthenogenesis (B) Parthenocarpy
(C) Apomixis (D) None

19. Centre of origin of rice is
(A) SW Asia (B) SE Asia
(C) South America (D) North America

20. The GOT is done for verification of
(A) Physical purity (B) Genetic purity
(C) Germination % (D) None

21. Flaring of square in cotton is due to
(A) Pink boll worm (B) Spotted boll worm
(C) American boll worm (D) All of the above

22. **Possible reasons for seed dormancy is**
(A) Presence of pathogens (B) Cracking of hulls
(C) Immature embryo (D) Green distoration

23. **Middle layer of fruit wall is called**
(A) Mesocarp (B) Mericarp
(C) Meristem (D) Micropyle

24. **In bhendi, production of foundation seed needs an isolation distance of**
(A) 100 metres (B) 50 metres
(C) 200 metres (D) 3 metres

25. **Seed coat is derived from**
(A) Testa (B) Embryo
(C) Endosperm (D) Nucellus

26. **In sunflower, production of foundation seed requires an isolation distance of (metres)**
(A) 400 (B) 800
(C) 200 (D) 100

27. **In wheat, production of foundation seed needs an isolation distance of**
(A) Three metres (B) Five metres
(C) Ten metres (D) Twenty metres

28. **Initial seed of an improved variety is called**
(A) Nucleus seed (B) Breeder seed
(C) Foundation seed (D) Certified seed

29. **Physical purity of 95% is permissible for the foundation and certified seeds of**
(A) Soybean (B) Groundnut
(C) Spinach (D) Carrot

30. **Physical purity of 98% is permissible in the crop**
(A) Groundnut (B) Rice
(C) Soybean (D) Spinach

31. **ISTA is located in ________**
(A) Germany (B) Denmark
(C) Netherlands (D) Switzerland

32. **Production of foundation seed in cotton requires an isolation distance of**
(A) 20 metres (B) 30 metres
(C) 50 metres (D) 75 metres

33. **In bhendi, production of certified seed needs an isolation distance of**
(A) 100 metres (B) 50 metres
(C) 200 metres (D) 3 metres

34. Perisperm is derived from
(A) Testa (B) Embryo
(C) Endosperm (D) Nucellus

35. In bittergourd, production of foundation seed requires an isolation distance of (metres)
(A) 1500 (B) 800
(C) 200 (D) 1000

36. In black gram, production of foundation seed needs an isolation distance of
(A) Three metres (B) Five metres
(C) Ten metres (D) Twenty metres

37. Seed of an improved variety is maintained by the breeder
(A) Nucleus seed (B) Breeder seed
(C) Foundation seed (D) Certified seed

38. Physical purity of 97% is permissible for the foundation and certified seeds of
(A) Soybean (B) Groundnut
(C) Spinach (D) Carrot

39. Physical purity of 98% is permissible in the crop
(A) Groundnut (B) Carrot
(C) Soybean (D) Spinach

40. Plant Breeders' Rights are operating in
(A) Germany (B) Denmark
(C) Netherlands (D) All of the above

41. Production of breeder seed in cotton requires an isolation distance of
(A) 20 metres (B) 30 metres
(C) 50 metres (D) 75 metres

42. A seed cut into two halves and immersed in 0.1% triphenyl tetrazolium chloride solution. Its viability is indicated by development of ________ colouration
(A) Yellow (B) Red
(C) Blue (D) Pink

43. Freshly liberated seeds of *Erianthus hiemalis* do not germinate due to
(A) Occurrence of impermeable seed coat
(B) Immaturity of embryo
(C) Occurrence of embryo damage
(D) Absence of growth hormones

44. Freshly shed seeds can germinate immediately under favourable conditions in
(A) Xanthium (B) *Pisum sativum*
(C) *Zea mays* (D) Both (B) and (C)

45. Common cause of seed and bud dormancy is presence of
(A) Ethylene (B) Cytokinins
(C) Abscisic acid (D) Both (B) and (C)

46. Seeds of tomato do not germinate in its pulp due to
(A) Presence of ferulic acid (B) Presence of excess salt
(C) Absence of oxygen (D) Presence of ABA

47. Seed dormancy may be due to
(A) Permeable seed coat (B) Hard impermeable seed coat
(C) Thin seed coat (D) Lack of reserve food

48. Wheat germ is
(A) Cotyledon (B) Endosperm
(C) Embryo (D) Ovule

49. The hormone which can break seed dormancy is
(A) Coumarin (B) Ferulic acid
(C) ABA (D) GA

50. Mechanical injuring of seed coat to break dormancy is called
(A) Scarification (B) Stratification
(C) Impaction (D) Compaction

51. Thiourea is used in overcoming seed dormancy by
(A) Chemical scarification (B) Counteracting inhibitors
(C) Inducing cell division (D) Develop osmotic pressure

52. The most important external factor for seed germination is
(A) Light (B) Soil
(C) Oxygen (D) Water

53. The first process which occurs when the seed is placed in the soil is
(A) Photosynthesis (B) Respiration
(C) Imbibition (D) Solubilisation of food

54. Seeds placed deep in the soil do not germinate because they are
(A) Unable to get sufficient oxygen
(B) Without sufficient food to bring the seedling the surface
(C) Under pressure of overlying soil layers
(D) Unable to get light

55. During germination, micropyle of seed takes part in
(A) Forming weak for emergence of radicle
(B) Absorption of water
(C) Passage of gases
(D) Leaching inhibitors

56. The plant in which hypocotyl stores food is
(A) Bertholletia (B) Nymphaea
(C) Cotton (D) Rice

57. A dicot albuminous seed is
(A) Pea (B) Castor
(C) Bean (D) Gram

58. A dicot exalbuminous seed is
(A) Pea (B) Castor
(C) Rice (D) Wheat

59. In maize, a tubular sheath covers the plumule. It is
(A) Fleshy (B) Leathery
(C) Papery (D) Green

60. In maize, a tubuler sheath covers the plumule. It is
(A) Coleoptile (B) Coleorhiza
(C) Mesocotyl (D) Scutellum

61. Aleurone is
(A) Outer layer of scutellum in contact with endosperm
(B) Layer of pericarp specialized in absorption of water
(C) Layer present in the ovule that guides pollen tube
(D) Layer present on the outside of endosperm and having protein grains

62. Seed of black pepper is
(A) Endospermic (B) Exalbuminous
(C) Acotyledonous (D) Perispermic

63. Coleorhiza is
(A) Covering of radicle
(B) Covering of plumule
(C) Plumule and rudimentary sheath
(D) Radicle and root cap

64. A spongy white bilobed structures present at the narrow tip of castor seed is
(A) Strophiole (B) Caruncle
(C) Raphe (D) Remains of chalaza

65. Seeds are required for fruit growth
(A) Throughout (B) Early phase
(C) Late phase (D) Mid phase

66. A seed which does not possess micropyle and hilum is
(A) Pea (B) Gram
(C) Castor (D) Maize

67. In Ricinus, the outer white papery covering of endosperm is
(A) Perisperm/tegmen (B) Tigellum
(C) Strophiole (D) Endosperm

68. The functions of PPV&FR is

1. Registration of new plant varieties, essentially derived varieties (EDV), extant varieties
2. Developing DUS (Distinctiveness, Uniformity and Stability) test guidelines for new plant species
3. Maintenance of National Gene Bank
4. Issuing of seed certification for foundation and certified seed.

(A) 1 alone correct
(B) 2 and 4 alone correct
(C) 1, 2 and 3 are correct
(D) All are correct

69. Find out the correct statement

1. Variety is defined by the expression of the characteristics resulting from a given genotype of that plant grouping
2. Variety is distinguished from any other plant grouping by expression of at least one of the said characteristics
3. Cultivate crops by cultivating the land himself

(A) 1 alone correct (B) 1 and 2 are correct
(C) 2 and 3 are correct (D) All are correct

70. Dicot endospermic seed is

(A) Castor (B) Fenugreek
(C) Both (A) and (B) (D) Bean

71. Winged seeds occur in

(A) Moringa (B) Ulmus
(C) Shorea (D) Hiptage

72. Which hormone is formed in leaves and helps in the blooming of the flowers?

(A) Traumatic (B) Auxin
(C) Florigen (D) None of the above

73. Which plant hormone is formed in injured cells by which the injury of plant is healed ?

(A) Florigens (B) Traumatic
(C) Abscisic acid (D) None of the above

74. Which of the following hormone help in the cell division and development in the presence of auxins?

(A) Ethylene (B) Cytokinin
(C) Auxin (D) Florigen

75. Which of the following is NOT a plant hormone?

(A) Corticosteroid (B) Brassinosteroid
(C) Polyamines (D) Salicylic acid

76. Which of the following secretes toxins during storage conditions of crop plants?

(A) *Aspergillus* (B) *Verticillium*
(C) *Fusarium* (D) *Colletotrichum*

77. What is bolting?
(A) Internode elongation (B) Rooting
(C) Shooting (D) Shoot apical meristem

78. At which humidity percentage should the cereal grains be stored ?
(A) 5 - 6% (B) 7 - 9%
(C) 10 - 15% (D) 8 - 10%

79. National Seed Corporation of India was established in the year
(A) 1961 (B) 1965
(C) 1963 (D) 1960

80. The process of breaking / softening the seed coat is called
(A) Caprification (B) Clarification
(C) Scarification (D) Saponification

81. Certification is not required for
(A) Breeder seed (B) Certified seed
(C) Foundation seed (D) Nucleus seed

82. Gladious is propagated by
(A) Corms (B) Bulbs
(C) Suckers (D) Buddings

83. Seed testing is done for
(A) C3 plants (B) C4 plants
(C) CAM plants (D) All the above

84. In sunflower, production of foundation seed requires an isolation distance of
(A) 400 metres (B) 800 metres
(C) 200 metres (D) 100 metres

85. Match list I with list II using the codes given below

List I	**List II**
(a) Bulb	1. Gladiolus
(b) Corm	2. Onion
(c) Rhizome	3. Potato
(d) Tuber	4. Turmeric

	(a)	(b)	(c)	(d)
(A)	1	2	4	3
(B)	2	1	4	3
(C)	1	2	3	4
(D)	2	1	3	4

86. What is the maximum permissible off types permitted for 99.5% genetic purity?
(A) 0.10% (B) 0.20%
(C) 0.30% (D) 0.50%

87. Which one of the following is not related to the functions of Central Seed Committee

(A) To advise central and state governments
(B) To certify seeds of any notified varieties
(C) To recommend standards for seed certification
(D) To recommend grow-out test standards

88. Which one of the following statement is not applicable for New Seed Policy, 1988

(A) Import of quality seed
(B) Export of quality seed
(C) Modernize plant quarantine
(D) Incentives to encourage domestic seed

89. Which of the following is not in line with the Seed Control Order, 1983

(A) Issue of licence to dealers
(B) Issue of test report to person
(C) Renewal of licence
(D) Appointing of licencing authority

90. The shedding tassels are removed and destroyed during the entire flowering phase in order to maintain 100% genetic purity in the case of ________ hybrid seed production fields

(A) Red gram (B) Cotton
(C) Sorghum (D) Maize

91. In tomato seed production plot, the seed certification officer is to observe ______ plants per count

(A) 50 plants (B) 100 plants
(C) 150 plants (D) 200 plants

92. In broadcasted seed crops like berseem and lucerne, how much area should be covered or field count?

(A) 85% (B) 75%
(C) 60-70% (D) 60%

93. How many field inspections can be taken up for verification in okra seed crop

(A) 4 (B) 3
(C) 2 (D) 1

94. Which one of the following method is suitable for seeds that are not extremely chaffy and do not bounce or roll?

(A) Spoon method (B) Random cup method
(C) Modified halving method (D) Hand halving method

95. When cereal grains germinate, the food reserve of the _____ are used first.

(A) Scutellum (B) Suspensor
(C) Aleurone layer (D) Endosperm

96. Match the following

(A) Tea	1. Recalcitant
(B) Cocoa	2. Capsule
(C) Cardamom	3. Berry
(D) Coffee	4. Orthodox

(A) 4, 2, 1, 3 (B) 2, 1, 3, 4
(C) 3, 1, 2, 4 (D) 4, 3, 2, 1

97. Match the following

Crop	**Explant**
(A) Banana	1. Meristem tip
(B) Rose	2. Suckers
(C) Anthurium	3. Shoot tip
(D) Carnation	4. Vegetative buds

(A) 4, 3, 2, 1 (B) 2, 3, 4, 1
(C) 1, 3, 4, 2 (D) 2, 4, 3, 1

98. Match the following

Crop	**Number of harvests**
(a) Cotton	1. 1 - 2 harvests
(b) Bhendi	2. 6 - 8 pickings
(c) Brinjal	3. 5 - 6 pickings
(d) Pearl millet	4. 4 - 5 pickings

(A) 4, 3, 2, 1 (B) 4, 2, 3, 1
(C) 3, 4, 2, 1 (D) 2, 4, 3, 1

99. Match the following

Crop	**Method of threshing**
(a) Rice	1. Stripping
(b) Groundnut	2. Flailing and pounding
(c) Greengram	3. Animal threshing
(d) Finger millets	4. Beating

(A) 2, 1, 4, 3 (B) 4, 2, 3, 1
(C) 1, 3, 4, 2 (D) 2, 4, 3, 1

100. Match the following

(a) Seed shape	1. Specific gravity separator
(b) Surface texture	2. Indented cylinder separator
(c) Seed length	3. Spiral separator
(d) Seed weight	4. Magnetic separator

(A) 4, 3, 2, 1 (B) 4, 2, 3, 1
(C) 1, 3, 4, 2 (D) 3, 4, 2, 1

Answer Keys

Q.No.	Ans.	Q.No.	Ans.	Q.No.	Ans.	Q.No.	Ans.	Q.No.	Ans.
1	D	21	B	41	C	61	D	81	A
2	C	22	C	42	C	62	D	82	A
3	A	23	A	43	B	63	A	83	D
4	B	24	C	44	D	64	B	84	A
5	B	25	A	45	C	65	B	85	B
6	A	26	A	46	A	66	D	86	D
7	C	27	A	47	B	67	A	87	B
8	D	28	A	48	C	68	C	88	B
9	A	29	D	49	D	69	B	89	B
10	B	30	B	50	A	70	C	90	D
11	C	31	D	51	B	71	A	91	B
12	A	32	C	52	D	72	C	92	D
13	B	33	C	53	C	73	B	93	B
14	B	34	D	54	A	74	B	94	B
15	A	35	A	55	B	75	A	95	A
16	C	36	B	56	A	76	A	96	B
17	C	37	A	57	B	77	A	97	B
18	B	38	B	58	A	78	D	98	C
19	B	39	C	59	C	79	C	99	A
20	B	40	D	60	A	80	C	100	D

MODEL TEST 10

1. **Which of the following is a parthenocarpic fruits?**
   (A) Orange (B) Papaya
   (C) Pomogranate (D) Apple
2. **Apomixis is**
   (A) Development of plant in darkness
   (B) Development of plants without fusion of gametes
   (C) Inability to perceive stimulus for flowering
   (D) Effect of low temperature on plant growth
3. **Adventitive polyembryony occurs in**
   (A) Poa (B) Brassica.
   (C) Allium (D) Citrus
4. **An orthotropous ovule is one in which micropyle and chalaza are**
   (A) In straight line of funicle (B) Parallel along with funicle
   (C) At right angles to funicle (D) Parallel along with ovule
5. **Anemophillous flowers have**
   (A) Sessile stigma (B) Small smooth stigma
   (C) Coloured flower (D) Large feathery stigma
6. **Anthesis is**
   (A) Opening of flower bud
   (B) Pollen mother cell undergoes meiosis
   (C) Dehiscence of anthers
   (D) Stigma becoming receptive
7. **Double fertilization involves**
   (A) Fusion of secondary nucleus with both male gametes
   (B) Fusion of polar nuclei with one male gamete, and egg and male gamete
   (C) Fusion of one polar nucleus with one male gamete
   (D) Any of above, depending on species
8. **Cauliflory is flowers produced**
   (A) On old dormant bud (B) On young bud
   (C) On the axil (D) On the branch
9. **Cleistogamous condition is present in**
   (A) *Brassica* (B) *Solanum tuberosum*
   (C) *Commelina* (D) *Allium cepa*
10. **Layering can be done on plants such as**
    (A) Bulbs (B) Bougainvillea
    (C) Sugar cane (D) Oleander

11. Artificial methods of reproduction do not include

(A) Rhizome (B) Cutting
(C) Layering (D) Budding

12. Vegetative propagation is another term for

(A) Asexual reproduction (B) Sexual reproduction
(C) Growing plants (D) Grafting

13. With an increase in seed size upon germination, dry mass of seeds

(A) Increases (B) Decreases
(C) Vanishes (D) Grows proportionally with seed

14. No matter how favorable environmental conditions are, seeds do not germinate if

(A) They have not settled
(B) The neighboring seeds do not germinate
(C) The seeds do not undergo a dormant or resting period
(D) The seeds are soaked in water

15. Dormant seeds are usually

(A) Succulent (B) Dry
(C) Wet (D) Soft

16. Germination of 101 years old seeds, was observed in

(A) Melon family (B) Arctic tundra lupine
(C) Moth mullein (D) Hesperidium

17. First step in seed germination is

(A) Passing through the gut of an animal
(B) Exploding along the sutures
(C) Absorption of water
(D) Taking in of O_2

18. Autochory occurs in

(A) *Geranium* (B) *Lotus*
(C) *Plumbago* (D) *Mango*

19. In censer mechanism, seed dispersal occurs through

(A) Birds
(B) Entangling of seeds to skin of roaming animals
(C) Shaking by wind
(D) Floating on water

20. Censer mechanism of seed dispersal occurs in

(A) *Bombax* (B) *Alstonia*
(C) *Moringa* (D) *Poppy*

21. Very small and light seeds which are dispersed by air are

(A) Dust seeds (B) Winged seeds
(C) Balloon seeds (D) Plumed seeds

22. Dust seeds occur in

(A) *Ulmus* (B) *Orchis*
(C) *Cinchona* (D) *Ailanthus*

23. Mendel discovered principles of inheritance because he
(A) Observed simultaneously all of the many characteristics in which the parents differed.
(B) Believed that the hereditary characteristics of two individuals became thoroughly blended in the offspring.
(C) Ignored all characteristics except a few markedly contrasting ones in which he studied.
(D) Studied only the offspring obtained from a single mating.

24. Which one of the following crop shows blonded seed ?
(A) Off-coloured seed (B) Blackened plumule
(C) Blonded seed (D) Whiptail

25. Which nutrient causes whiptail in crucifers
(A) Molybdenum (B) Calcium
(C) Boron (D) Nitrogen

26. A strand of DNA with the sequence A A C T T G will have a complimentary strand with the following sequence
(A) CCAGGT (B) AACTTG
(C) TTCAAG (D) TTGAAC

27. A pedigree chart shows
(A) The genotypic ratios of the offspring.
(B) The types of gametes produced by the parents.
(C) The pattern of inheritance of a specific gene.
(D) None of these

28. Gene technology involves the modification of
(A) NDA (B) DNA
(C) ADP (D) DAA

29. When an organism is modified using gene technology, the organism is then referred to as a
(A) Generally modified organism
(B) Very modified organism
(C) Genetically modified organism
(D) Gently modified organism

30. What was the first genetically modified crop to be sold in supermarkets in the United States?
(A) Flavr Savr tomato (B) Herculex I corn
(C) Knockout corn (D) Endless Summer tomato

31. In what year were GMO seeds first available commercially in the United States?
(A) 1984 (B) 1989
(C) 1996 (D) 1998

32. What is the most popular GMO crop cultivated in the United States?
(A) Corn (B) Soybean
(C) Cotton (D) Squash

33. Seed drying is very important to maintain its

(A) Viability and vigour (B) Protein content
(C) Oil content (D) Chemical composition

34. The equipment used to apply chemicals that involves suspension and wettable powder treatment material in water is

(A) Slurry treaters (B) Direct treaters
(C) Panogen treaters (D) Mist-o-matic treaters

35. Fungi and bacteria usually enters through

(A) Stomata (B) Wounds
(C) Hydathodes (D) Insect puntyre

36. Which element plays a key role in chlorphyll formation

(A) Magnesium (B) Molybdenum
(C) Chlorine (D) Boron

37. Which one of the following wheat type grown in major area in India.

(A) *Triticale* (B) *Triticum durum*
(C) *T. dicoccum* (D) *T. aestivum*

38. The use of legislative restriction to control pests is called

(A) Plant enclosure (B) Plant protection
(C) Plant law (D) Plant quarantine

39. A herbicide applied to the soil after planting a crop is called a

(A) Postplant herbicide (B) Pre emergence herbicide
(C) Postemergence herbicide (D) Preplant herbicide

40. What are the units of LD_{50}?

(A) gram/gram of animal body weight
(B) kilogram/gram of animal body weight
(C) milligrams/ kilogram of animal body weight
(D) milligrams/gram of animal body weight

41. In an IPM program, ______ control is considered first?

(A) Chemical (B) Mechanical
(C) Cultural (D) Biological

42. The technique used for hybrid seed production in maize is

(A) Emasculation and dusting (B) Three line breeding syatem
(C) Detasseling (D) Chemical emasculants

43. Which of the following test is used to predict the genetic purity of seed ?

(A) Potassium nitrate test (B) Ferric chloride test
(C) Grow out test (D) Seed germination test

44. Unwanted plants growing from seed that remains in the field from a previous crop is termed as

(A) Objectionble weed plants (B) Volunteer plants
(C) Pollen shedders (D) Rogues

45. The level of natural outcrossing in groundnut is

(A) 5% (B) 10%
(C) 15% (D) Negligible

46. The Tarai Development Corporation was established in the year

(A) 1965 (B) 1966
(C) 1963 (D) 1969

47. Match the following:

(a) Spiral separator — 1. Length of the seed
(b) Disc separator — 2. Shape and surface texture
(c) Magnetic separator — 3. Shape or rolling ability
(d) Cylinder separator — 4. Surface texture and stickiness of seed

(A) 1 2 3 4 (B) 3 1 4 2
(C) 2 3 1 4 (D) 4 1 3 2

48. Reproduction which does not involve gametic union is

(A) Para sexual reproduction (B) Apomixis
(C) Parthenogenesis (D) Agamospermy

49. Nucellar embryo is

(A) Amphimictic haploid (B) Amphimictic diploid
(C) Apomictic haploid (D) Apomictic diploid

50. In parthenocarpy, which levels are high in ovaries

(A) Gibberellins (B) Cytokinins
(C) Auxins (D) All of these

51. Bananas are seedless because they

(A) Are reproduce asexually (B) Are triploid
(C) Are sprayed with hormone (D) Are parthenocarpic

52. Formation of fruits without fertilization is known as

(A) Parthenogenesis (B) Parthenocarpy
(C) Polyembryony (D) Polygamy

53. In hiptage, anemochory occurs because of

(A) Winged seeds (B) Balloon fruits
(C) Winged fruits (D) Plumed fruits

54. In anemochorous tecoma, the modification present is

(A) Dust seeds (B) Winged seeds
(C) Winged fruits (D) Plumed seeds

55. A fleshy fruit dispersed by inflated persistent calyx is

(A) *Bombax* (B) *Albizzia*
(C) *Physalis* (D) *Naravelia*

56. Spacing for sorghum seed production is

(A) 40 × 10 cm (B) 45 × 10 cm
(C) 45 × 15 cm (D) 45 × 25 cm

57. In redgram hybrid seed production, female row consists of
(A) 50% sterile and 50% fertile
(B) 75% sterile and 25% fertile
(C) 75% fertile and 25% sterile
(D) All the plants sterile

58. There are three basic requirements for protection of a variety under PBR
(A) Stability (B) Distinctiveness
(C) Uniformity (D) All the above

59. Apomixis is a form of
(A) Parthenocarpy (B) Vernalization
(C) Parthenogenesis (D) None of these

60. In which hybrid seed production, hand rubbing with muslin cloth is followed.
(A) Sunflower (B) Maize
(C) Pearl Millet (D) Redgram

61. Which one of the following is a stress test
(A) Brick gravel test (B) Seed size
(C) First count (D) Tetrazolium test

62. An allele is
(A) Another word for a gene
(B) A homozygous genotype
(C) A heterozygous genotype
(D) One of several possible forms of a gene

63. Where Central Rice Research Institute is located?
(A) Calcutta (B) Cuttack
(C) Karnal (D) New Delhi

64. In which hybrid seed production, pod driving is practiced?
(A) Rice (B) Maize
(C) Sunflower (D) Cotton

65. What is the seed rate for male and female in maize hybrid seed production?
(A) 5:10 kg/ha (B) 10:10 kg/ha
(C) 15:10 kg/ha (D) 10:15 kg/ha

66. What is the suitable season for cumbu hybrid seed production?
(A) November-December (B) January-Febrary
(C) March-April (D) May – June

67. There is an increase in _________ under drought conditions.
(A) ABA content (B) Ethylene level
(C) Proline content (D) All of these

68. Brassinosteroids are lipids synthesized from
(A) Cholesterol (B) Stigmasterol
(C) Campesterol (D) Ergosterol

69. Which one of the following hormone is used for root formation?
(A) GA (B) Auxins
(C) ABA (D) Cytokinins

70. A simple one seeded fruit with pericarp fused with the seed coat is known as
(A) Caryopsis (B) Achene
(C) Nut (D) Cypsela

71. Which one of the following is the maximum standard for inert matter of groundnut?
(A) 1% (B) 2%
(C) 3% (D) 4%

72. An isolation distance of 200m is followed in certified seed production of
(A) Brinjal (B) Cucurbits
(C) Chilli (D) Onion mother bulb

73. A minimum of 70% is the germination standard for
(A) Cauliflower (B) Carrot
(C) Cabbage (D) Curry leaf

74. Which one of the following method is adopted for tomato seed extraction?
(A) Wet seed extraction (B) Dry seed extraction
(C) Acid / Alkali method (D) Manual extraction

75. ______________ is the preconditioning method adopted for castor for TZ test.
(A) Bisect longitudinally (B) Bisect laterally
(C) Remove seed coat (D) No conditioning

76. What is the SMR for pigeon pea
(A) 1:49 (B) 1:150
(C) 1:120 (D) 1:200

77. What is the optimum stage of harvesting of cauliflower seed
(A) 7 weeks after anthesis (B) 6 weeks after anthesis
(C) 5 weeks after anthesis (D) 4 weeks after anthesis

78. What is the size of working sample for purity analysis for okra
(A) 140 g (B) 70 g
(C) 120 g (D) 60 g

79. Phylloclade is
(A) Underground modification of stem
(B) Sub-aerial modification of stem
(C) Aerial modification of stem
(D) Root modification

80. In one litre buffer solution, how many grams of TZ salt should be mixed to prepare 1% solution.
(A) 0.1 g (B) 1 g
(C) 10 g (D) 100 g

81. Which of the following is a plant whose seeds are dispersed by wind?
(A) Elm (B) Oak
(C) Willow (D) Cottonwood

82. An example for combined dormancy
(A) Pine (B) Teak
(C) Maple (D) Ficus

83. Which one of the following is a treatment for breaking physical dormancy
(A) Scarification (B) Stratification
(C) Warm stratification (D) Cold stratification

84. When pre-moistened seeds are exposed to 20-25°C, the treatment is known as ________
(A) Scarification (B) Stratification
(C) Warm stratification (D) Cold stratification

85. When pre-moistened seeds are subjected to 3-5°C, the treatment is called as
(A) Scarification (B) Stratification
(C) Abrasion (D) Warm temperature

86. The condition in which the seed can germinate upon absorption of water is known as
(A) Dormancy (B) Quiescence
(C) Primary dormancy (D) Secondary dormancy

87. ________ is an example for homogamy
(A) Maize (B) Pearl millet
(C) Mirabilis (D) Sapota

88. The definitive nucleus on fertilization gives to
(A) Secondary (B) Zygote
(C) Oospore (D) Primary endospermic nucleus

89. ________ is the maximum standard for inert matter in the case of bhendi.
(A) 1% (B) 2%
(C) 3% (D) 4%

90. Which one of the following is the minimum standard for germination of foundation seed of redgram?

(A) 75% (B) 80%

(C) 85% (D) 90%

91. An example of a fruit that disperses seeds by splitting along one or more edges of the carpel is

(A) Samara (B) Follicle

(C) Hesperidium (D) Drupe

92. Which from the following statement is not in accordance to the duties of seed inspectors

(A) To take seed samples of any notified variety from a person who is selling seeds

(B) To enter into the sale point where he suspects any offence

(C) To send the test results to the person from whom he received seed sample

(D) To examine any record or register found in the place where he suspects offence

93. Plants that grown upon other plants, but do not absorb food from them are called

(A) Parasites (B) Epiphytes

(C) Saprophytes (D) Symbionts

94. When an F1 is crossed with the recessive homozygote parent the cross is known as

(A) Back cross (B) Top cross

(C) Test cross (D) Single cross

95. What colour tag is used for International seed sample certificate?

(A) White (B) Green

(C) Blue (D) Orange

96. Match the following

List I	List II
(a) Mango	1. Panicle
(b) Coconut	2. Cyme
(c) Cauliflower	3. Spadix
(d) Clove	4. Raceme

(A) 4, 1, 2, 3 (B) 2, 3, 4, 1

(C) 1, 3, 4, 2 (D) 3, 2, 4, 1

97. Match the following

List I	List II
(a) Allogamy	1. Self pollination
(b) Autogamy	2. Cross pollination
(c) Cleistogamy	3. Vallisneria
(d) Hydrophily	4. Arachis

(A) 3, 2, 4, 1 (B) 2, 3, 4, 1
(C) 4, 1, 2, 3 (D) 2, 1, 4, 3

98. Match the following

List I	List 2
(a) Mitophily	1. Moths
(b) Palenophily	2. Dicliny
(c) Unisexuality	3. Hybrid cotton
(d) Xenoganay	4. Bees

(A) 4, 1, 2, 3 (B) 3, 2, 4, 1
(C) 2, 3, 4, 1 (D) 2, 1, 4, 3

99. Match the following

List I	List II
(a) Entomophily	1. Castor
(b) Waspophily	2. Bombax
(c) Ornithophily	3. Maize
(d) Psycophily	4. Coriander

(A) 2, 3, 4, 1 (B) 3, 2, 4, 1
(C) 3, 4, 2, 1 (D) 2, 1, 4, 3

100. Match the following

List I	List II
(a) Drupe	1. Grapes
(b) Berry	2. Coffee
(c) Schizocarp	3. Bhendi
(d) Capsule	4. Carrot

(A) 2, 1, 4, 3 (B) 1, 2, 3, 4
(C) 3, 4, 2, 1 (D) 4, 3, 1, 2

Answer Keys

Q.No.	Ans.	Q.No.	Ans.	Q.No.	Ans.	Q.No.	Ans.	Q.No.	Ans.
1	D	21	A	41	C	61	A	81	B
2	B	22	B	42	C	62	D	82	B
3	D	23	B	43	C	63	A	83	A
4	A	24	C	44	B	64	A	84	C
5	D	25	A	45	D	65	A	85	D
6	C	26	D	46	D	66	C	86	B
7	B	27	C	47	B	67	D	87	C
8	D	28	B	48	B	68	C	88	C
9	C	29	C	49	D	69	B	89	A
10	B	30	A	50	C	70	A	90	A
11	A	31	D	51	B	71	D	91	B
12	A	32	B	52	D	72	C	92	C
13	B	33	A	53	C	73	C	93	B
14	C	34	A	54	C	74	C	94	C
15	B	35	A	55	C	75	C	95	C
16	C	36	A	56	C	76	B	96	C
17	C	37	D	57	A	77	A	97	D
18	A	38	D	58	D	78	A	98	A
19	B	39	A	59	C	79	C	99	C
20	D	40	C	60	A	80	C	100	A

MODEL TEST 11

1. **Maturity of stamens and pistils at different time is called as**
   (A) Cleistogamy (B) Apogamy
   (C) Dioecy (D) Dicliny

2. **The study of pollen is termed as**
   (A) Cartilinology (B) Palentology
   (C) Ornithology (D) Palenology

3. **The stalk of the pistil between stigma and ovary is called as**
   (A) Style (B) Peduncle
   (C) Ligule (D) Auricle

4. **An example for parthenocarpy fruit is**
   (A) Mango (B) Seedless watermelon
   (C) Sapota (D) Papaya

5. **Starch is stored in**
   (A) Nucleus
   (B) Mitochondria
   (C) Golgi body
   (D) Amyloplast and Cotyledon

6. **Caryopsis is the seed type of**
   (A) Paddy (B) Sunflower
   (C) Groundnut (D) Carrot

7. **The hard seed coat can be removed by**
   (A) Acid scarification (B) After ripening
   (C) Drying (D) Stratification

8. **Inhibitor present in cabbage seed is**
   (A) ABA (B) Coumarin
   (C) Hydroxy urea (D) Methyl urea

9. **Precocious germination in seeds of fleshy fruits is prevented by**
   (A) More moisture (B) ABA content
   (C) Osmotic stress (D) Desiccation stress

10. **GA synthesis and signaling dominate the transition of**
    (A) Dormancy (B) Sensitivity
    (C) Germination (D) Receptivity

11. **The phenomenon of heterosis of the progeny over better parent is**
    (A) Heterosis (B) Heterobeltiosis
    (C) Relative heterosis (D) Standard heterosis

12. Top cross is otherwise known as

(A) Inbred variety cross (B) Single cross
(C) Double cross (D) Triple cross

13. What is the seed renewal period for hybrid?

(A) Every year (B) Once in 2 years
(C) Once in 3 years (D) Once in 4 years

14. A rogue is

(A) Normal plant (B) Crop plant
(C) Off type plant (D) A line

15. Five generation system of seed multiplication can be followed in

(A) Self pollinated crops
(B) Often cross pollinated crops
(C) Cross pollinated crops
(D) Hybrids

16. Effect of pollen grains on the maternal tissues or the effect of foreign pollen on the endosperm colouration in maize is called as

(A) Paramutation (B) Metaxenia
(C) Penetrance (D) Regulon

17. Rain during maturation promoted the production of following seed in pulses

(A) Off coloured seed (B) Mottled seed
(C) Dimbled seed (D) Blackened seed

18. SRR = X / Y × 100. In this formula Y is

(A) Quantity of seed produced (B) Quantity of seed sown
(C) Quantity of seed required (D) Quantity of seed supplied

19. Presence of B line in A line of paddy hybrid seed production field is

(A) Pollen shedder (B) Off-type
(C) Rogue (D) Weed

20. Staggered sowing is practiced for

(A) Good field establishment (B) Tolerance of pest and diseases
(C) Uniform flowering (D) Easy processing

21. Split husk commonly occurs in

(A) Barley (B) Wheat
(C) Oats (D) Paddy

22. Cob sorting in maize is done to identify

(A) Diseased cob (B) Matured cob
(C) Big sized cob (D) Off type cob

23. Planting ratio for redgram hybrid seed production is

(A) 6 : 2 (B) 4 :2
(C) 4 :1 (D) 3 :1

24. Pulses seed set could be improved by given foliar spray during flowering with

(A) DAP (B) $ZnSO_4$
(C) Borax (D) Urea

25. Important nutrient for groundnut seed production is

(A) Copper (B) Magnesium
(C) Manganese (D) Calcium

26. In sunflower, pollination is brought about by

(A) Honey bees (B) Bungle bees
(C) Wind (D) Butterflies

27. The maximum percentage of pollen shedders, that is permitted in case of cotton hybrid is

(A) 0.05 % (B) 1.0 %
(C) 5.0 % (D) 10 %

28. In jute seed production, the first rogueing is to be carried out when the plants are

(A) 30 days old (B) 25 – 35 days old
(C) 30 – 40 days old (D) 35 – 45 days old

29. Planting ratio for tomato hybrid seed production is

(A) 5 : 1 (B) 4 :1
(C) 3 :1 (D) 2 :1

30. Germination standards for foundation and certified seed production of bhendi is

(A) 80% (B) 70%
(C) 75% (D) 65%

31. Isolation distance for snake gourd foundation seed production is

(A) 100m (B) 1000m
(C) 1500m (D) 500m

32. Which cucurbitaceous vegetables having a large volume of fibrous portion when fully ripe

(A) Snake gourd (B) Bitter gourd
(C) Sponge gourd (D) Bottle gourd

33. Formation of an elongated stem or seed stalk that occurs generally during the second season of growth in case of biennial plants is called as

(A) Pome (B) Achene
(C) Bolt (D) Caryopsis

34. How the carrot certified seed production is done?

(A) Seed to seed method (B) Root to seed method
(C) Stump to seed method (D) Bud to seed method

35. Daincha seed crop attain physiological maturity at

(A) 40 DAA (B) 45 DAA
(C) 50 DAA (D) 60 DAA

36. In alfalfa, flowers are usually tripped for efficient pollination by

(A) Wind (B) Rain
(C) Wasps (D) Honey bees

37. Viviparous seed present in

(A) Cocoa (B) Nutmeg
(C) Clove (D) Cinnamon

38. Seed moisture content of coffee seeds at the time of berry stage is about

(A) 25 % (B) 50%
(C) 75 % (D) 100%

39. Haulm cutting in potato is adopted to avoid

(A) Viral disease (B) Bacterial disease
(C) Genetic contamination (D) Physical contamination

40. Number of tissue culture plants required for planting an acre of sugarcane is

(A) 2000 (B) 3000
(C) 4000 (D) 5000

41. Types of apomixis in apple is

(A) Facultative (B) Obligatory
(C) Non recurrent (D) Recurrent

42. Mango seed germination is improved by

(A) Seed husking
(B) Soaking in GA_3
(C) Soaking in IAA
(D) Seed husking and soaking in GA_3

43. Pollination in orchid is carried out by

(A) Insects (B) Beetle
(C) Wind (D) Water

44. Exogenous and endogenous dormancy is exhibited in which of the following crop

(A) Rose (B) Orchid
(C) Dahlia (D) Petunia

45. Which cells can be induced to reproduce an entire plant

(A) Meristematic cells (B) Parenchytmatus cells
(C) Colenchymatus cells (D) Parietal cells

46. In micro propagation, the explants are surface sterilized by

(A) Potassium permanganate (B) Sodium thiosulphate
(C) Calcium chloride (D) Sodium hypochlorite

47. The male sterile parent is crossed with which one of the following line to produce hybrid seed

(A) A line (B) B line
(C) R line (D) C line

48. Which one of the following formula is used to calculate heterobeltiois

(A) $\frac{F-BP}{BP} \times 100$ (B) $\frac{F1-MP}{MP} \times 100$

(C) $\frac{F-CV}{CV} \times 100$ (D) $\frac{F1-CP}{CP} \times 100$

49. Female sterility in sorghum is deficiency symptom of

(A) Fe (B) Zn

(C) Boron (D) Mn

50. Designated objectionable weed in wheat seed crop

(A) *Chicorum intybus* (B) *Cucurbita* sp

(C) *Convolvulus arvensis* (D) *Chicorum* sp

51. A seed with fairly unequal cotyledons

(A) Oak (B) Jack fruit

(C) Castor (D) Tamarind

52. Vivipary is the characteristic feature of

(A) Hydrophytes (B) Xerophytes

(C) Halophytes (D) Mesophytes

53. When the concentration of the soil solutes is low, the absorption of water is

(A) Remain normal (B) Increased

(C) Decreased (D) Stopped

54. Second process that occurs during germination is

(A) Imbibition (B) Rupturing of seed coat

(C) Gibberellin synthesis (D) Amylase synthesis

55. Which of the following exhibits the high rate of respiration?

(A) Growing shoot apex (B) Germination of seed

(C) Root tip (D) Leaf bud

56. Potato tubers are larger in plants grown at Darjeeling than grown in AP, this is because there is the rate of

(A) Respiration is high

(B) Respiration is low

(C) Photosynthesis is high

(D) Protein synthesis is high

57. An albuminous seed showing hypogeal germination

(A) Castor (B) Bean

(C) Gram (D) Maize

58. A gas required for germination of pea seed is

(A) Nitrogen (B) Oxygen

(C) Hydrogen (D) Water vapour

59. Newly synthesized DNA in radicle tips is associated with

(A) Nuclear membrane (B) Endosperm

(C) Cytoplasm (D) Protoplasm

60. Which may inhibit the germination by inhibiting the DNA polymerase ?

(A) GA_3 (B) ABA

(C) Kinetin (D) Auxin

61. The example for manual scarification is

(A) *Eucalyptus* (B) *Casuarina*

(C) *Terminalia* (D) *Pterocarpus*

62. Which acid is used for acid scarification?

(A) Dilute KOH (B) Commerical KOH

(C) Diltue H_2SO_4 (D) Commercial H_2SO_4

63. In seed bank, the seeds are preserved at ______ temperature.

(A) 18°C (B) -18°C

(C) 19°C (D) -19°C

64. __________ is the physiological changes associated with seed deterioration?

(A) Reduced respiration (B) Mitochondrial fragmentation

(C) Seed colour changes (D) Delayed germination

65. The short lived seeds are known as ______________.

(A) Megabiotic seed (B) Macrobiotic seed

(C) Mesobiotic seed (D) Microbiotic seed

66. Which one of the following treatment is not a storage treatment?

(A) Halogenation (B) Antioxidant treatment

(C) Fumigation (D) Magnetic treatment

67. Washing method is not suitable for health testing of which one of the following fungi?

(A) Smut of sorghum (B) Karnal bunt of wheat

(C) Loose smut of wheat (D) Downy mildew of soybean

68. ________% NaOH is used in NaOH seed soak method for testing of storage fungi.

(A) 0.1% (B) 0.2%

(C) 0.5% (D) 1.0%

69. Which one of the following fungi penetrates through the wounds and infects the seed internally?

(A) *Ustilage tritici* (B) *Pseudomonas syringae*

(C) *Ascochyta pinodella* (D) *Ustilago nuda*

70. ________ is the pathogen that stick to the surface of the seed?

(A) *Ustilage tritici* (B) *Pseudomonas syringae*

(C) *Ascochyta pinodella* (D) *Ustilago nuda*

71. Loss of seed due to birds is ________

(A) 10-24% (B) 5-15%

(C) 12-34% (D) 22-42%

72. How many chapters are there in PPV & FR?

(A) 9 (B) 11
(C) 10 (D) 12

73. Which one of the following is not related to notable features of PPV & FR?

(A) Protection of plant varieties
(B) Protection of farmers right
(C) Promotion of competitive plant
(D) Ensures supply of good quality seed breeding

74. In which year SFCI (State Farm Corporation of India) was framed?

(A) 1969 (B) 1968
(C) 1967 (D) 1966

75. In which year, the first cotton hybrid was released?

(A) 1969 (B) 1970
(C) 1971 (D) 1972

76. Which one of the following material is not issued with phytosanitary certificate

(A) Tissue culture plants (B) Sexually propagated material
(C) Imported seed (D) Diseased seed

77. In which five year plan, Central Seed Certification board was established?

(A) Fifth (B) Sixth
(C) Seventh (D) Eighth

78. Which one of the following organisation is conserving the wheat germplasm?

(A) IRRI (B) CIMMYT
(C) IITA (D) CIP

79. The chemical used for cryopreservation is ________

(A) Liquid hydrogen (B) Liquid oxygen
(C) Liquid nitrogen (D) Liquid carbon

80. Which country is maintaining the Svalbard Global Seed Vault?

(A) USA (B) Norway
(C) London (D) Germany

81. Biological effect of lipid peroxidation

(A) Mitochondria dysfunction (B) Enzyme inactivation
(C) Membrane pertuberance (C) All the above

82. What kind of damage occur in chromosome of deteriorating seed?

(A) Rubbing (B) Deteriorating
(C) Aberration (D) Rupturing

83. The internal factor that influence the seed deterioration is
(A) Seed moisture (B) RH
(C) Temperature (D) Stress

84. What flower leant a gene to rice to synthesize the production of beta carotene in "golden rice" ?
(A) Tulip (B) Daffodil
(C) Rose (D) Sunflower

85. Transformation method of plants and animals in which plants and animals are given shocks is known as
(A) Microinjection (B) Genome breeding
(C) Electroporation (D) Genome engineering

86. In what year did Europeans put a stop to GMOs by refusing to sell it in their markets or food stands?
(A) 1986 (B) 1999
(C) 2001 (D) 2005

87. +ve photoblastic seeds are
(A) Nigella (B) Silene
(C) Nemophills (D) Arraranutum

88. Dispersal of seeds through explosive mechanism occurs in
(A) Tecoma (B) Sonchus
(C) Squirting cucumber (D) Poppy

89. Which fungal disease spreads by seed and flowers?
(A) Loose smut of wheat
(B) Corn smut
(C) Covered smut of barley
(D) Soft rot of potato.

90. Transgenic plants
(A) Contain foreign genes in their cells
(B) Are used to produce human antibodies
(C) Both (A) and (B)
(D) Are plants that differ in geographical locations

91. Which of the following is used in biowar ?
(A) A pathogen
(B) Toxin from a pathogen
(C) A delivery system for the bioweapon agent
(D) All of these

92. Number of countries growing GMO crops is/are
(A) 28 (B) 25
(C) 30 (D) 35

93. The first transgenic food crop to be allowed for cultivation in the country
(A) Brinjal (B) Mustard
(C) Corn (D) Soybean

94. What do you mean by conservation?

(A) It means the protection, preservation, management, or restoration of wildlife and natural resources such as forests and water.
(B) It means an ethic of resource use, allocation and protection.
(C) Only (A)
(D) Both (A) & (B)

95. Which one of the following is not used for *ex-situ* plant conservation?

(A) Seed banks
(B) Field gene banks
(C) Shifting cultivation
(D) Botanical gardens

96. Match the following

(A) Boerner divider	1. Small seeds
(B) Gamet divider	2. Gravitational force
(C) Spoon method	3. Chaffy seeds
(D) Hand halving method	4. Centrifugal force

(A) 2, 4, 1, 3　　(B) 2, 4, 3, 1
(C) 1, 3, 4, 2　　(D) 4, 2, 3, 1

97. Match the following

Seed sample (wt)	**Decimal**
(a) < 1 g	1. 3
(b) > 1 g to < 10 g	2. 1
(c) > 10 g to < 100 g	3. 4
(d) > 100 g to < 1000 g	4. 2

(A) 4, 3, 2, 1　　(B) 3, 1, 4, 2
(C) 2, 3, 4, 1　　(D) 2, 4, 3, 1

98. Match the following

Preparation step for staining	**Example**
(a) Bisect longitudinally	1. Maize
(b) Bisect laterally	2. Dicots
(c) Remove seed coat	3. Small seeded grasses
(d) Conditioning only	4. Large seeded legumes

(A) 3, 1, 2, 4　　(B) 4, 2, 3, 1
(C) 1, 3, 2, 4　　(D) 2, 3, 4, 1

99. Match the following

(a) Top of paper	1. Legume seeds
(b) Hard seed	2. Medium sized seeds
(c) Between paper	3. Large sized seeds
(d) Sand media	4. Light requiring seeds

(A) 4, 3, 2, 1　　(B) 4, 1, 2, 3
(C) 1, 2, 4, 3　　(D) 2, 4, 3, 1

100. Match the following

(A)	Mechanical damage	1.	Conductivity test
(B)	Dead tissue identification	2.	$FeCl_3$ test
(C)	Deterioration process	3.	Tetrazolium test
(D)	Germinable seed	4.	X-ray analysis

(A) 4, 3, 2, 1 (B) 4, 2, 1, 3
(C) 2, 4, 1, 3 (D) 2, 4, 3, 1

Answer Keys

Q.No.	Ans.	Q.No.	Ans.	Q.No.	Ans.	Q.No.	Ans.	Q.No.	Ans.
1	B	21	D	41	D	61	C	81	D
2	D	22	D	42	D	62	D	82	C
3	A	23	D	43	A	63	B	83	A
4	B	24	A	44	A	64	D	84	B
5	D	25	D	45	A	65	D	85	C
6	A	26	A	46	D	66	D	86	B
7	A	27	A	47	A	67	C	87	D
8	C	28	C	48	A	68	B	88	C
9	B	29	A	49	A	69	B	89	A
10	C	30	D	50	C	70	C	90	C
11	B	31	B	51	C	71	C	91	D
12	A	32	C	52	B	72	B	92	A
13	A	33	C	53	B	73	D	93	B
14	C	34	B	54	A	74	A	94	D
15	A	35	A	55	B	75	C	95	C
16	B	36	D	56	B	76	C	96	A
17	A	37	A	57	D	77	D	97	B
18	A	38	B	58	B	78	B	98	C
19	A	39	D	59	A	79	C	99	A
20	C	40	A	60	B	80	B	100	B

MODEL TEST 12

1. **Decortication is used for**
   (A) Maize (B) Sunflower
   (C) Bhendi (D) Groundnut
2. **Wet extraction is commonly used in**
   (A) Rice (B) Bittergourd
   (C) Chillies (D) Onion
3. **Extraction of seeds or stones by removal of the fleshy part (pulp) of fruits like berries and drupes is called as**
   (A) Depulping (B) Decantitation
   (C) Pulping (D) Pulp decantation
4. **The process of removing large size particles in seed processing is**
   (A) Pre-cleaning (B) Basic cleaning
   (C) Scalping (D) Upgrading
5. **The temperature limit for artificial air drying**
   (A) 110°F (B) 100°C
   (C) 65°F (D) 110°C
6. **Moisture content of seeds in equilibrium with atmospheric humidity at a given temperature is termed as**
   (A) Inequilibrium moisture content
   (B) Standard moisture content
   (C) Graded moisture content
   (D) Equilibrium moisture content
7. **Flat and round seeds can be separated by**
   (A) Specific gravity separator
   (B) Spiral separator
   (C) Magnetic separator
   (D) Indent cylinder separator
8. **What is the bottom screen size for rajmash seed processing?**
   (A) 8/64" (B) 10/64"
   (C) 12/64" (D) 16/64"
9. **Commonly used fungicide for seed treatment is**
   (A) Furadan (B) Metasystox
   (C) Carbendazim (D) Atrazine
10. **What is the dosage of neem leaf powder for seed treatment?**
    (A) 5 g/kg of seed (B) 10 g/kg of seed
    (C) 15 g/kg of seed (D) 20 g/kg of seed

11. Which one of the following crop can be given with seed hardening treatment using 2% dihydrogen phosphate?

(A) Cotton (B) Safflower
(C) Blackgram (D) Sorghum

12. Halopriming involves

(A) Water soaking (B) Osmotic solution soaking
(C) Salt solution soaking (D) Mixing with inorganic carrier

13. Up to which moisture level, the seeds are soaked while fortification of seeds?

(A) 15 - 20% (B) 20 - 25%
(C) 25 - 30% (D) 30 - 35%

14. The use of *Rhizobium* in seed pelleting is known as

(A) Inoculant pelleting (B) Herbicide pelleting
(C) Protective pelleting (D) Organic pelleting

15. What is the optimum dose of polykote for seed coating?

(A) 2 g/kg of seed (B) 3 g/kg of seed
(C) 4 g/kg of seed (D) 5 g/kg of seed

16. Controlled hydration is

(A) Fortification (B) Hardening
(C) Priming (D) Infusion

17. The treatment of seeds with an osmotic solution i.e., polyethylene glycol to initiate germination and then dried to get more uniform and rapid germination of certain vegetable seeds is called as

(A) Priming (B) Stratification
(C) Hardening (D) Fortification

18. A seed treatment to enhance the market value

(A) Seed priming (B) Polykote treatment
(C) Seed dipping (D) Seed fortification

19. PEG is commonly used in

(A) Halopriming (B) Osmopriming
(C) Hydropriming (D) Solid matrix priming

20. Procedure by which individual seeds are provided with an envelope of adhesive material containing e.g. nutrients, microsymbiont inoculants and / or pesticides is known as

(A) Grading (B) Stacking
(C) Crushing (D) Pelleting

21. For which crop, the specific gravity separator separated seeds are in positive association with vigour

(A) Cotton (B) Sunflower
(C) Casuarina (D) Cashewnut

22. Air screen cleaner used for

(A) Scalping and grading (B) Grading
(C) Separating (D) Cleaning

23. **Magnetic separator uses**
(A) Calcium chloride powder (B) Iron powder
(C) Magnesium chloride powder (D) Chalk powder

24. **For conveying seeds vertically, the following equipment is used**
(A) Screw conveyor (B) Bucket elevator
(C) Belt conveyor (D) Vertical conveyor

25. **What is the concentration of fast green solution used for testing seed coat cracking?**
(A) 0.25% (B) 0.50%
(C) 0.75% (D) 1.0%

26. **National Seeds Corporation was established to develop Indian Seed Industry in the year**
(A) 1968 (B) 1963
(C) 1967 (D) 1964

27. **Orange certificate is issued by**
(A) ISTA (B) OECD
(C) AOSA (D) UPOV

28. **Breeder seed tag colour is**
(A) Golden yellow (B) Azhur blue
(C) White (D) Opal green

29. **Genetic contamination is caused by**
(A) Other crop plants (B) Weed plants
(C) Diseased plants (D) Off type plants

30. **Pearson's square law is used in**
(A) Seed sampling (B) Seed blending
(C) Seed dividing (D) Seed mixing

31. **Number of plants counted per single count during field inspection of bhendi, brinjal and chilli is**
(A) 100 plants (B) 200 plants
(C) 500 plants (D) 1000 plants

32. **If the area of the field is 6.1 to 8 acres, how many counts you will make in field inspection of a seed crop?**
(A) 5 (B) 6
(C) 7 (D) 8

33. **Which one of the following crop needs 1000 g submitted sample for grow out test**
(A) Wheat (B) Paddy
(C) Maize (D) Sorghum

34. **The maximum number of ODV/kg of seed permitted for maize composite is**
(A) 5 (B) 10
(C) 15 (D) 20

35. The optimum germination temperature for a wide range of seeds is between

(A) 25 and 30°C (B) 30 and 35°C
(C) 10 and 15°C (D) 35 and 40°C

36. In purity analysis, seeds less than half size is considered as

(A) Other crop seed (B) Inert matter
(C) Pure seed (D) Weed seed

37. At the end of the germination test period, neither hard nor fresh and have not produced seedlings, are classified as

(A) Dead seeds (B) Obsolete seed
(C) Recalcitrant seed (D) Calcitrant seed

38. If the seeds are kept in 1-5 containers, what is the sampling intensity?

(A) Each container, from atleast 5 positions
(B) Not less than 5 containers
(C) Each container, from atleast 6 positions
(D) Atleast one container in every 5 positions

39. In which containers, the seeds should be packed and sent to STL for moisture estimation?

(A) Moisture pervious container
(B) Moisture proof container
(C) Cloth bag
(D) Gunny bag

40. Chemical used in quick viability test is

(A) Potassium chloride
(B) Gibberellic acid
(C) Sodium chloride
(D) 2,3-5 Triphenyl tetrazolium chloride

41. The concept of seed vigour originally given by _____ in 1911.

(A) Hiltner Inssen (B) Perry
(C) Isley (D) Moore

42. Which one of the following is the physiological test?

(A) Brick gravel test (B) Seed size
(C) First count (D) Tetrazolium test

43. Which one of the following method cannot be used for quantitative assay of transgenes?

(A) RT-PCR (B) Southern
(C) PCR (D) Q-PCR

44. Phenol test is recommended for varietal identification of

(A) Paddy (B) Groundnut
(C) Blackgram (D) Brinjal

45. The head quarters of ISTA is located in

(A) Zurich (Switzerland) (B) Connecticut (USA)
(C) Tharanxndt (Germany) (D) Melbourne (Australia)

46. The seeds having short life
(A) Mango (B) Greengram
(C) Rice (D) Pearl millet

47. The first segment of the storage period represents the time interval between
(A) Physiological maturity and harvest maturity
(B) Morphological maturity and physiological maturity
(C) Harvest and processing
(D) Harvest maturity and processing

48. For safe storage, the required moisture content of rice seed is below
(A) 8 % (B) 10 %
(C) 14 % (D) 17 %

49. At what RH, the fungi and bacteria are very active during storage?
(A) 62% (B) 75%
(C) < 75% (D) >75%

50. Phospholipids plays a vital role in
(A) Photosynthesis (B) Cell-permeability
(C) Transpiration (D) Water absorption

51. Seed plants use meiosis to produce
(A) Pollens and seeds (B) Spores and pollens
(C) Seeds and fruits (D) Microspores and megaspores

52. Plant development includes all of the following except
(A) Seed formation
(B) Tissue formation
(C) Formation of embryo and suspensor
(D) Deteriorating cotyledons during germination

53. The development of female gametophyte from the megaspore mother cell
(A) Megasporogenesis (B) Megagametogenesis
(C) Double fertilization (D) Ovule primordial

54. Starch is major food reserve in
(A) Cereal seeds (B) Oil seeds
(C) Pulse seeds (D) Cotton seeds

55. Which of the following tissues makes up the majority of the interior of the plant?
(A) Epidermal tissue (B) Ground tissue
(C) Meristem (D) Phloem

56. Which of the following tissues forms the outer protective covering of the plant?
(A) Epidermal tissue (B) Ground tissue
(C) Meristem (D) Vascular system

57. Inhibitor present in cabbage seeds
(A) ABA (B) Coumarin
(C) Hydroxy urea (D) Methyl urea

58. Cell cycle activity is suppressed in
(A) Developing seed (B) Dormant seed
(C) Maturing seed (D) Germinating seed

59. Which of the following plant hormones is incorrectly paired with its function?
(A) Auxin – responsible for apical dominance
(B) Abscisic acid - Regulates the rate of respiration
(C) Cytokinin – Delays Senescence
(D) Gibberellins - Promote bud and seed dormancy

60. Dormancy in seeds and buds can be broken by application of which hormone?
(A) Abscisic acid (B) Auxins
(C) Cytokinin (D) Gibberellins

61. Which of the following doesn't exhibit seed dormancy?
(A) *Phaseolus* (B) *Xanthium*
(C) *Rhizophora* (D) *Casssia*

62. Primary growth in plant is initiated by
(A) Apical meristem (B) Vascular cambium
(C) Lateral meristems (D) Ground tissue

63. Lipolysis occurs in
(A) Cytosol (B) Oil bodies
(C) Endoplasmic reticulum (D) Mitochondria

64. Which one of the following is a morphological change associated with seed deterioration ?
(A) Reduced respiration (B) Mitochondrial fragmentation
(C) Seed colour changes (D) Delayed germination

65. _______________ is medium lived seed.
(A) Megabiotic seed (B) Macrobiotic seed
(C) Mesobiotic seed (D) Microbiotic seed

66. What is the optimum moisture content of seed for undertaking fumigation?
(A) Around 14% (B) Below 14%
(C) Above 14% (D) More than 14%

67. The karnal bunt infected seeds become ________ in colour and that can be detected easily.
(A) White (B) Brown
(C) Black (D) Red

68. Which one of the following can be used for confirming the presence of karnal bunt?
(A) Teliospore (B) Sclerotia
(C) Pycnidia (D) Conidia

69. Which pathogen enters the seed through the wall of ovary?

(A) *Fusarium oxysporum*
(B) *Xanthomonas compestris*
(C) *Ustilaga tritici*
(D) *Rhizoctonia solani*

70. _____________ penetrates through the stigma and infects the seed internally.

(A) *Fusarium oxysporum*
(B) *Xanthomonas compestris*
(C) *Ustilaga tritici*
(D) *Rhizoctonia solani*

71. Which one of the following storage pest can infest the seed in the field before harvest?

(A) Pulse beetle (B) Rice weevil
(C) Larger grain borer (D) Confused flour beetle

72. Which one of the following variant limit is considered unstable for cereal while testing stability of variety?

(A) 8 variants in 150 ear to row progeny
(B) More than 8 variants in 150 ear to row progeny
(C) Less than 8 variants in 150 ear to row progeny
(D) Not less than 8 variants in 150 ear to row progeny

73. Which one of the following is not pertaining to the criteria for protection / registration of varieties?

(A) Notable (B) Distinctiveness
(C) Uniformity (D) Stability

74. First private seed company was

(A) Monsanto (B) Namdhari
(C) Sutton & Sons (D) Takii

75. The first symbolic importance of seed started after the report of Famine Commission in :

(A) 1881 (B) 1890
(C) 1891 (D) 1892

76. Under PPV & FR Act, compulsory license is issued to the registered variety after expiry of _________ years from the date of registration?

(A) 1 (B) 2
(C) 3 (D) 4

77. Which one of the following organisation follows single window system of marketing?

(A) Private seed producer (B) National Seeds Corporation
(C) Co-operative sector (D) Agriculture Department

78. In which year, the Seed Review Team was launched?

(A) 1969 (B) 1968
(C) 1967 (D) 1966

79. Which one of the following is the old consensus of seed deterioration?

(A) Impaired transcription

(B) Production of free radicals causing incomplete enzyme synthesis damages to the cell

(C) Hindering of mitochondrial replication

(D) Lesser ATP synthesis

80. DMSO (Dimethyl sulfoxide) is used as

(A) Gelling agent (B) Alkaylating agent

(C) Chelating agent (D) Cryoprotectant

81. Name the first naturally occurring cytokinin.

(A) Neoxanthin (B) Xanthoxin

(C) Zeatin (D) Isopentenyl adenine

82. Which of the following plant hormone is responsible for seed germination?

(A) Auxin (B) Gibberellin

(C) Ethylene (D) Abscisic acid

83. Which company is the largest biotech company in the world, also responsible for the majority of the GMOs in our food supply ?

(A) Kroger (B) Nestle

(C) Monsanto (D) Nike

84. What are the two of the top 4 GMO ?

(A) Wheat and barley (B) Soybean and corn

(C) Tomato and pepper (D) Kale and spinach

85. The controversy regarding the use of Bt corn is that it

(A) Is potentially harmful to monarch butterflies

(B) Is a potential allergen to humans

(C) Both (A) and (B)

(D) Can contaminate groundwater

86. What do you mean by conservation of Biodiversity?

(A) It envisions preservation, maintenance, sustainable use (conservation), recovery and enhancement of the components of biological diversity.

(B) It is the process of protecting an endangered plant or animal species in its natural habitat, either by protecting or cleaning up the habitat itself or by defending the species from predators.

(C) It is the proper management of the biosphere by human beings in such a way that it gives maximum benefits for the present generation and also develops its potential so as to meet the needs of the future generations.

(D) All of the above

87. Consider the statement and choose the correct answer.

(i) Seed is a ripened ovule

(ii) Genetic purity of breeder seed is 99.5%

(iii) Seed is a carrier of food supply

(A) (i) alone is correct (B) (ii) alone is correct

(C) (iii) alone is correct (D) (i) & (ii) are correct

88. Match List I with List II and choose the correct options given below

List I	List II
a. Genetic purity	1. Free from pests & diseases
b. Physical purity	2. High germination & vigour
c. Physiological quality	3. Physical contamination
d. Health quality	4. Genetic contamination

	a	b	c	d
(A)	4	3	2	1
(B)	3	4	2	1
(C)	3	2	4	1
(D)	1	2	3	4

89. In which crop, blonded seed occurs?

(A) Black gram (B) Green gram

(C) Tomato (D) Peas

90. Vivipary is seed germination

(A) In strong light

(B) Without involving cotyledons

(C) With cotyledons coming above ground

(D) While contained inside the fruit

91. Gibberellins promote

(A) Seed germination (B) Seed dormancy

(C) Leaf fall (D) Root elongation

92. Vivipary occurs in

(A) Mangrove plants

(B) Alpine plants /orchids

(C) Tropical plants/sea weeds

(D) Desert plants/ Vallisneria

93. Match the following:

(A) Calcium	1. Blotching
(B) Boron	2. Blossom end rot
(C) Molybdenum	3. Hollow heart
(D) Potassium	4. Whiptail

(A) 4, 2, 1, 3 (B) 3, 2, 4, 1

(C) 4, 3, 2, 1 (D) 2, 3, 4, 1

94. The 17th essential element added for plant growth is

(A) Cobalt (B) Molybdenum

(C) Nickel (D) Magnesium

95. Match the following:

(a) Nitrogen	1. Integral part of amino acids
(b) Phosphorus	2. Building blocks of protein
(c) Sulphur	3. Energy storage
(d) Magnesium	4. Key element of chlorophyll production

(A) 2, 3, 1, 4 (B) 4, 3, 2, 1
(C) 3, 4, 2, 1 (D) 1, 3, 2, 4

96. Match the following

(a) Quick viability test	1. Baskin and Dalhousie
(b) Brick gravel test	2. Lakon
(c) Accelerated ageing test	3. Hilter
(d) Speed of germination	4. Maequer

(A) 2, 3, 1, 4 (B) 3, 2, 4, 1
(C) 1, 3, 2, 4 (D) 2, 4, 3, 1

97. Match the following

(a) Cool germination test	1. Barium chloride
(b) Cold test	2. Soybean
(c) Indoxy acetate test	3. Corn
(d) Cotrast method	4. Cotton

(A) 4, 3, 2, 1 (B) 4, 2, 1, 3
(C) 2, 3, 4, 1 (D) 2, 4, 3, 1

98. Match the following

Vigour test	**Example**
(a) Physical test	1. Vigour index
(b) Performance test	2. Seed size
(c) Biochemical test	3. Water sensitivity test
(d) Stress test	4. Tetrazolium test

(A) 4, 3, 2, 1 (B) 4, 2, 3, 1
(C) 2, 3, 4, 1 (D) 2, 1, 4, 3

99. Match the following

Varietal identification	**Test**
(a) Morphological test	1. Coleoptile colour
(b) Physiological test	2. RFLP
(c) Chemical test	3. Phenol test
(d) Molecular test	4. Seed weight

(A) 4, 3, 2, 1 (B) 4, 2, 3, 1
(C) 2, 3, 4, 1 (D) 4, 1, 3, 2

100. Match the following

(a) Hypocotyl colour	1. Wheat
(b) Cotyledon colour	2. Beetroot
(c) Disease resistance test	3. Brassica
(d) 2, 4-D test	4. Maize

(A) 2, 3, 1, 4 (B) 4, 2, 1, 3
(C) 1, 3, 2, 4 (D) 2, 4, 3, 1

Answer Keys

Q.No.	Ans.	Q.No.	Ans.	Q.No.	Ans.	Q.No.	Ans.	Q.No.	Ans.
1	D	21	B	41	A	61	A	81	C
2	B	22	A	42	C	62	B	82	B
3	A	23	B	43	C	63	B	83	C
4	C	24	B	44	A	64	C	84	B
5	A	25	D	45	A	65	C	85	C
6	D	26	B	46	A	66	Both A & B	86	D
7	B	27	A	47	A	67	C	87	A
8	C	28	A	48	C	68	A	88	A
9	C	29	D	49	C	69	D	89	D
10	B	30	B	50	B	70	C	90	D
11	D	31	A	51	D	71	A	91	A
12	C	32	D	52	D	72	B	92	A
13	B	33	C	53	B	73	A	93	D
14	A	34	D	54	A	74	C	94	C
15	B	35	A	55	B	75	A	95	A
16	C	36	B	56	A	76	C	96	A
17	A	37	A	57	C	77	B	97	A
18	B	38	A	58	B	78	B	98	D
19	B	39	B	59	D	79	A	99	D
20	D	40	D	60	D	80	D	100	A

MODEL TEST 13

1. **Which one of the following is the physiological deterioration of seed?**
   (A) Loss of weight (B) Loss of germination
   (C) Protein decrease (D) Insect damage

2. **For long term storage of seeds, the suitable seed storage container is**
   (A) Gada cloth bag (B) Gunny bag
   (C) Polythene bag (D) Paper bag

3. **Polyethylene bags of <300 gauge thickness allows**
   (A) Entry of water in the form of vapour and liquid
   (B) Entry of water in the form of liquid and not in vapour
   (C) Entry of water in the form of vapour and not in liquid
   (D) Entry of water not in the form of both liquid and vapour

4. **Foundation of seed warehouse should be extended above the ground level with the height of**
   (A) 50 cm (B) 70 m
   (C) 90 cm (D) 110 cm

5. **In refrigerated storage, how much temperature should be maintained for 60% relative humidity?**
   (A) 57°F (B) 67°F
   (C) 70°F (D) 62°F

6. **Which one of the following material is a conductor of heat in seed storage building?**
   (A) Wood (B) Glass wool
   (C) Stone (D) Steel

7. **The largest germplasm storage bank in the world is situated in**
   (A) Colorado, USA (B) Tokyo, Japan
   (C) New Delhi, India (D) Moscow, Russia

8. **Which institute is responsible for the germplasm collections of paddy?**
   (A) IRRI (B) CIMMYT
   (C) ITTA (D) AURDC

9. **Which activity of the cell is stopped during cryopreservation?**
   (A) Metabolic (B) Biochemical
   (C) Enzyme (D) Respiration

10. Which one of the following is a fumigant?

(A) Methyl bromide (B) Methyl cyanite
(C) Methyl dematon (D) Methyl ethyl chloride

11. Which one of the disease causing pathogen enters into seed through pollen tube?

(A) Whip tail (B) Grain smut
(C) Loose smut (D) Wilt

12. Designated disease for wheat is

(A) Kernel bunt (B) Ergot
(C) Black rot (D) Kernel smut

13. Designated disease for rice is

(A) Kernel bunt (B) Ergot
(C) Black rot (D) Kernel smut

14. Test for pathogen infection in seed is known as

(A) Health test (B) DUS test
(C) Patent test (D) Vigour test

15. The toxin produced by *Aspergillus flavus*

(A) Aflatoxin (B) Ochratoxin
(C) Fumerosin (D) Zeatin

16. The concentration of 2,4-D used for 2,4-D blotter test

(A) 100 ppm (B) 1000 ppm
(C) 0.2 % (D) 2 %

17. PCNB media is most useful media to detect

(A) *Penicillium* (B) *Phythium*
(C) *Aspergillus* (D) *Fusarium* spp

18. Solar treatment is used for

(A) Stem rust (B) Loose smut
(C) Powdery mildew (D) Blast

19. A certificate from a recognized plant quarantine service that indicates that a sample is substantially free from disease or pests is called as

(A) DUS certificate (B) Certification
(C) Quarantine (D) Phytosanitary certificate

20. ____________ is a seed borne viral disease

(A) Cabbage yellow mosaic (B) Tobacco mosaic
(C) Common bean mosaic (D) Bhendi mosaic

21. The post harvest losses of grains in our country as estimated by Government of India is

(A) 10% (B) 9%
(C) 8% (D) 9.5%

22. Important storage grain pest for pulse is

(A) Rice moth (B) Grain moth
(C) Pulse beetle (D) Khapra beetle

23. An example for external feeder is
(A) Rice weevil
(B) Granery weevil
(C) Seed beetles
(D) Red flour beetle

24. The symptom of pulse beetle attack is
(A) Circular holes
(B) Irregular holes
(C) Damaged seed
(D) Webbing

25. Irregular holes is the symptom of
(A) Red flour beetle
(B) Pulse beetle
(C) Lesser grain borer
(D) Mites

26. For grain cum seed storage of pulses, seeds are treated with
(A) Activated clay
(B) Thiram
(C) Captan
(D) Bavistin

27. Infestation of legume seeds by bruchids results in complete loss of viability within _____ months of storage
(A) 3-5
(B) 5-8
(C) 4-6
(D) 2-4

28. What is the minimum seed certification standard for nematode infection in sweet potato certified seed crop?
(A) 2%
(B) 6%
(C) 8%
(D) 10%

29. The infection of nematode leads to discolouraton of testa in groundnut by
(A) *Anguina* spp
(B) *Ditylenchus* spp
(C) *Aphelenchoides* spp
(D) *Heterodera* spp

30. CPRI is located in
(A) Wellington
(B) Shimla
(C) Kashmir
(D) Ooty

31. Indian Society of Seed Technology was formed in
(A) 1969
(B) 1972
(C) 1975
(D) 1971

32. Activities of state seed certification agencies is co-ordianated by
(A) NSC
(B) Central Seed Committee
(C) Central Seed Certification Board
(D) SFCI

33. The seed year is
(A) 1961
(B) 1975
(C) 1980
(D) 1981

34. First AICCIP was introduced in
(A) Maize
(B) Paddy
(C) Groundnut
(D) Redgram

35. Syngenta belongs to which country?
(A) USA
(B) UK
(C) Germany
(D) Switzerland

36. Where Bharwale foundation is located?
(A) Jalna (B) Hyderabad
(C) Bangalore (D) Coimbatore

37. Seed multiplication ratio for groundnut is
(A) 1:40 (B) 1:20
(C) 1:10 (D) 1:8

38. In which year seed review team was started?
(A) 1958 (B) 1968
(C) 1978 (D) 1988

39. In which year, the first SAU was started?
(A) 1957 (B) 1958
(C) 1959 (D) 1960

40. The duration of protection of varieties of rice and maize under PPV & FRA, 2001 is
(A) 15 years (B) 18 years
(C) 6 years (D) 9 years

41. Compulsory licence will be granted after the expiry of
(A) 3 years from the date of issue of certificate
(B) 4 years from the date of issue of certificate
(C) 5 years from the date of issue of certificate
(D) 2 years from the date of issue of certificate

42. Chapter VI under PPV&FR Act, 2001 deals with
(A) Compulsory license
(B) Surrender / revocation of certificate
(C) Appellate Tribunal
(D) Farmers' Rights

43. How many members are present in PPV&FR Authority ?
(A) 12 (B) 15
(C) 18 (D) 21

44. Which one of the following award instituted under National Gene Fund of PPV&FR Authority ?
(A) Plant Genetic Resources Conservation Award
(B) Best Farmer / Community in Genome Conservation
(C) Plant Genome Saviour Community Award
(D) Genetic Resources Conservation Award

45. DUS test fee for vegetables is
(A) Rs.30,000 (B) Rs. 20,000
(C) Rs.15,000 (D) Rs. 39,000

46. Plant Variety Journal of India is published by
(A) PPV & FRA (B) PBR
(C) ICAR (D) IARI

47. Where UPOV is located?
(A) New York (B) Paris
(C) London (D) Geneva

48. In which crop, the first GMO plant was developed?
(A) Tapioca (B) Turmeric
(C) Tobacco (D) Tamarind

49. In which crop, GMO plant has been developed to improve fuel product
(A) Neem (B) Pungam
(C) Jatropha (D) Paradise tree

50. The rate of fatty acid oxidation is increased by
(A) Phospholipids (B) Glycolipids
(C) Aminolipids (D) Sulpholipids

51. DNA synthesis occurs in
(A) Quiescent seed (B) Dry seed
(C) Imbibed seed (D) Treated seed

52. The enzyme carboxylase requires a mineral activator. It is
(A) Mg (B) Mo
(C) Fe (D) Mn

53. In monocots, gibberellins are released from the
(A) Scutellum (B) Aleurone layer
(C) Endosperm (D) Embryo

54. Method that reders the seed coat permeable to water so that embryo expansion is not physically retarted is called as
(A) Scarification (B) Denudation
(C) Vernalization (D) Stratification

55. Ability of the seed to give rise to a seedling is called as
(A) Germination (B) Viability
(C) Dormancy (D) Vigour

56. The presence of cotyledons above the soil surface is refers to
(A) Epigeal (B) Hypogeal
(C) Vivipary (D) Insitu

57. Humidity in atmosphere decreases rate of
(A) Glycolysis (B) Photosynthesis
(C) Transpiration (D) Growth

58. What is the name given to the food storage tissue that surrounds the embryo in a dicot seed?
(A) Coleorhiza (B) Plumule
(C) Radicle (D) Cotyledons

59. Extreme impermeability of water through the seed coat leads to
(A) Innate dormancy (B) Seed coat dormancy
(C) Induced dormancy (D) Embryo dormancy

60. DNA synthesis is associated with
(A) Post germinative stage (B) Development
(C) Initial phase of germination (D) Pre germination stage

61. Pine seed carbohydrate is stored in

(A) Radicle (B) Cotyledon

(C) Mega gametophyte (D) Perisperm

62. The long lived seed is called as __________________

(A) Megabiotic seed (B) Macrobiotic seed

(C) Mesobiotic seed (D) Microbiotic seed

63. What is the moisture content of sand in which the seeds are conditioned and dried?

(A) 0-5% (B) 5-10%

(C) 10-15% (D) 15-20%

64. What is the duration of alternating light and dark that is given for blotter method of seed health testing?

(A) 6 h (B) 12 h

(C) 18 h (D) 24 h

65. In agar method of seed health testing, the plated seeds are incubated for ________________ days

(A) 5-7 (B) 4-6

(C) 3-5 (D) 2-4

66. Which one of the following fungi is responsible for non systemic infection of episperm?

(A) *Sclerospora* (B) *Colletotrichum truncatum*

(C) *Neovossia indica* (D) *Corynebacterium michiganense*

67. What is the designated disease of paddy?

(A) Grain smut (B) Head smut

(C) Loose smut (D) Bunt

68. What is the tolerance limit of offtypes for self pollinated crops while testing uniformity of variety?

(A) 1% (B) 2%

(C) 3% (D) 4%

69. How many sections are provided in PPV & FRA?

(A) 97 (B) 98

(C) 99 (D) 100

70. Which one of the following was established by GOI and state governments on recommendation of Royal Commission on Agriculture?

(A) Research Institutes (B) Seed Law Enforcement

(C) Seed Testing Laboratories (D) Seed Certification

71. In which year, the private vegetable seed companies had developed seed production facilities?

(A) 1935 (B) 1940

(C) 1945 (D) 1950

72. In which year, first phase of National Seeds Programme was launched?
(A) 1973-74 (B) 1974-75
(C) 1975-76 (D) 1976-77

73. Give an example for multiple window system of marketing?
(A) NSC
(B) Agriculture Department
(C) Horticulture Department
(D) Agriculture Marketing Department

74. When Association of Official Seed Analysts was started?
(A) 1908 (B) 1907
(C) 1906 (D) 1905

75. Which one of the following is not related to hydration and dehydration treatment?
(A) Given only to stored seeds
(B) Given to low and medium vigour non-leguminous seeds
(C) Given to fresh seeds
(D) Avoid direct soaking of leguminous seed

76. Which one of the following is a bleaching powder?
(A) Calcium chloride
(B) Calcium carbonate
(C) Calcium bicarbonate
(D) Calcium oxychloride

77. Which one of the following seed is a recalcitrant seed?
(A) Guava (B) Pomegranate
(C) Grapes (D) Mango

78. For every decrease of ____________ in storage temperature, the seed storage life doubled.
(A) 5°F (B) 10°F
(C) 15°F (D) 20°F

79. At what height, the seed bags are stacked in storage godown?
(A) 6-8 tires (B) 7-9 tires
(C) 5-7 tires (D) 4-6 tires

80. At what seed moisture range, the field fungi activity is excess for their growth?
(A) 24-25% (B) 23-24%
(C) 25-26% (D) 26-27%

81. Which one of the following is not relevant to pure seed fraction?
(A) All botanical varieties of that kind / species
(B) Broken seeds, if the size is > ¾ of the original size.
(C) Broken seeds, if the size is > ½ of the original size.
(D) Immature, shrivelled seeds.

82. Which of the following is not a valid classification of weeds in a seed purity test?
(A) Common weed (B) Noxious weed
(C) Annual weed (D) Restricted

83. An decrease in germination had a good correlation with
(A) Increase in protein (B) Increase in electrolyte leakage
(C) Increase in fat (D) Increase in carbohydrate

84. Which one of the enzyme decreases during seed deterioration
(A) α-amylase (B) β-amylase
(C) Oxidase (D) Peroxidase

85. During ageing, the lipids are broken into
(A) Free fatty acid (B) Acid value
(B) Iodine value (D) Amino acids

86. Seed longevity is a
(A) Qualitative trait (B) Quantitative trait
(C) Dominant (D) Recessive

87. Which substance is more dangerous for seed deterioration as per new consensus?
(A) Free radicle (B) Free radical
(C) Anion (D) Cation

88. Which one of the following is the physiological manifestation of seed deterioration?
(A) Decreased seed weight
(B) Decreased respiration
(C) Decreased abnormal seedling
(D) Delayed germination

89. Which of the following is a maintainer line?
(A) A line (B) B line
(C) R line (D) M line

90. Which one of the following is a top cross hybrid?
(A) A x R = F_1 (B) (A × B) × (C × D) = F_1
(C) (A × B) × C = F_1 (D) (A × B) × OPV = F_1

91. The new varieties of plants are produced by
(A) Introduction and mutation (B) Selection and hybridisation
(C) Mutation and selection (D) Selection and introduction

92. Heterosis over the better parent is called as
(A) Relative heterosis (B) Standard heterosis
(C) Pseudoheterosis (D) Heterbeltiosis

93. If a character is cytoplasmically controlled then
(A) The progeny resembles female parent
(B) The progeny resembles male parent
(C) It is segregation of 3:1 is observed in F_1
(D) 15:1 segregation in F_1

94. If a dihybrid without genic interaction is selfed the following ratio is obtained in F_2

(A) 1:1:1:1 (B) 9:3:3:1
(C) 9:7 (D) 9:6:1

95. Find out the incorrect statement of PPV&FRA.

1. The Act covers all categories of plants, except *microorganisms*
2. Apart from the breeders and others involved in developing new varieties, the Act recognizes farmers too.
3. For registration of a plant variety, the variety should possess with description for Novelty, Distinctiveness, Uniformity and Stability (NDUS).
4. The PPV&FR Authority shall starts receiving applications for Registration of Varieties of 12 notified crop species from 21st May, 2007.

(A) 1 alone incorrect (B) 2 & 3 are correct
(C) 2 alone incorrect (D) All are correct

96. Match the following

(a) Phenol test — 1. Soybean
(b) Peroxidase test — 2. Garden pea
(c) Sodium hydroxide — 3. Oats
(d) Potassium dichromate test — 4. Wheat

(A) 4, 3, 1, 2 (B) 4, 2, 3, 1
(C) 3, 1, 4, 2 (D) 2, 3, 4, 1

97. Match the following

(A) KOH beach test — 1. Millet
(B) Ferrous sulphate colour test — 2. Sorghum
(C) HCl test — 3. Paddy
(D) Iodine test — 4. Cucumber

(A) 4, 3, 2, 1 (B) 2, 3, 4, 1
(C) 1, 3, 2, 4 (D) 2, 4, 1, 3

98. (i) Orthodox seed is insensitive to low temperature
(ii) Recalcitrant seed is sensitive to low temperature
(iii) Example of moisture pervious container is polyethylene bag
(iv) Example of moisture impervious container is paper bag

(A) (i) and (iii) are correct (B) (ii) and (iii) are correct
(C) (i) and (ii) are correct (D) All are correct

99. Match the following

(A) Morphological changes — 1. DNA degradation
(B) Biochemical changes — 2. Discolouration
(C) Genetic changes — 3. Reduced seedling growth
(D) Physiological changes — 4. Loss of enzymes activity

(A) 4, 3, 1, 2 (B) 4, 2, 3, 1
(C) 1, 3, 2, 4 (D) 2, 4, 1, 3

100. Match the following

Pre storage treatments	**Minimum germination (%)**
(A) Halogenation	1. Vitamin E
(B) Antioxidant treatment	2. Aluminium phosphide
(C) Seed protectants	3. Bavistin
(D) Seed fumigation	4. Calcium oxychloride

(A) 4, 1, 3, 2 (B) 4, 2, 1, 3
(C) 1, 2, 4, 3 (D) 4, 2, 3, 1

Answer Keys

Q.No.	Ans.	Q.No.	Ans.	Q.No.	Ans.	Q.No.	Ans.	Q.No.	Ans.
1	B	21	D	41	A	61	C	81	B
2	C	22	C	42	D	62	B	82	C
3	C	23	D	43	B	63	B	83	B
4	C	24	A	44	C	64	B	84	D
5	D	25	C	45	B	65	A	85	D
6	D	26	A	46	A	66	C	86	B
7	A	27	D	47	D	67	D	87	B
8	A	28	D	48	C	68	A	88	D
9	A	29	B	49	C	69	A	89	B
10	A	30	B	50	A	70	A	90	D
11	C	31	D	51	C	71	C	91	B
12	A	32	C	52	A	72	C	92	D
13	D	33	A	53	A	73	D	93	A
14	A	34	A	54	A	74	A	94	B
15	A	35	D	55	A	75	C	95	D
16	C	36	B	56	A	76	D	96	C
17	C	37	D	57	C	77	D	97	B
18	B	38	B	58	D	78	B	98	C
19	D	39	C	59	B	79	A	99	D
20	D	40	A	60	A	80	A	100	A

MODEL TEST 14

1. Synthetic seeds are

(A) Artificially synthesized seeds
(B) Somatic embryos encapsulated in suitable matrix
(C) Seeds of plants modified genetically
(D) None of these

2. Somatic embryoids are

(A) Identical with zygotic embryos and without seed coats
(B) Identical with zygotic embryos and with seed coats
(C) Non-identical with zygotic embryos and without seed coats
(D) Non-identical with zygotic embryos and with seed coats

3. The production of high quality and uniform embryos has been limited to only

(A) Carrot (B) Alfalfa
(C) Both (A) and (B) (D) Sandalwood

4. The preserved embryoids are termed as

(A) Synthetic seeds (B) Semi-synthetic seeds
(C) Natural seeds (D) Fermented seeds

5. Encapsulation is necessary to produce and protect synthetic seeds. The encapsulation is carried out by various types of hydrogels, which are

(A) Soluble in water
(B) Soluble in organic solvents
(C) Insoluble in water
(D) Insoluble in organic solvents

6. In drosophila (fruit flies), eye colour is sex-linked and red eye colour is dominant to white eye colour. Which of the following are not possible in a cross between a red-eyed male and a heterozygous female?

(A) Red-eyed male (B) White-eyed male
(C) Carrier female (D) Homozygous white-eyed female

7. Which statement concerning a pair of alleles for a gene controlling a single characteristic in humans is true?

(A) Both genes come from the father
(B) Both genes come from the mother
(C) One gene comes from the mother and one gene comes from the father
(D) The genes come randomly in pairs from either the mother or father

8. Which of the following factors could lead to variations in the offspring of asexually reproducing organisms?

(A) Crossing over
(B) Fertilization
(C) Mutations
(D) Independent assortment

9. Genetic traits of seeds are noted as follows:

L = long, l = short
W = wrinkled, w = smooth
Y = yellow, y = white
R = ribbed, r = grooved

Which of the following is the genotype for a short, wrinkled, yellow, grooved seed?

(A) llWwyyrr (B) LLWWyYRr
(C) LlWwYYRr (D) llWwYYrr

10. Mendel discovered principles of inheritance because he:

(A) Observed simultaneously all of the many characteristics in which the parents differed.
(B) Believed that the hereditary characteristics of two individuals became thoroughly blended in the offspring.
(C) Ignored all characteristics except a few markedly contrasting ones in which he studied.
(D) Studied only the offspring obtained from a single mating.

11. Carriers of the colour-blindness trait include:

(A) Men who are heterozygous for the trait
(B) Men who are homozygous for the trait.
(C) Women who are heterozygous for the trait
(D) Women who are homozygous for the trait.

12. Which one of the following statement is wrong?

(A) Intine is made up of cellulose and pectin
(B) Vegetative cell is larger than generative cell
(C) Pollen grains in some plants remain viable for months
(D) When pollen is shed at two-celled stage, double fertilization does not take place

13. Vernalisation stimulates flowering in

(A) Carrot (B) Turmeric
(C) Elephant yam (D) Ginger

14. Ornithophily refers to the pollination by which of the following:

(A) Snails (B) Birds
(C) Insects (D) Air

15. Ovule is attached to the placenta of ovary wall by:

(A) Funicle (B) Raphae
(C) Hilum (D) Testa

16. Plants with ovaries having only one or a few ovules are generally pollinated by

(A) Birds (B) Butterflies
(C) Bees (D) Wind

17. Hot water treatment is used to break dormancy in

(A) Lentil (B) Cherry
(C) Cotton (D) None

18. Mak – 12 is a variety of

(A) Maize (B) Medicago
(C) Bt cotton (D) Tomato

19. Difference between seed and grain is

(A) Seed processing (B) Genetic purity
(C) Seed quality (D) Cost

20. Fanning mill is

(A) Air screen cleaner (B) Hammer mill
(C) Blower (D) Rice mill

21. National nematode collection of India is located at

(A) New Delhi (B) Dharwad
(C) Anand (D) Bangalore

22. Phenotype refers to the ____ of an individual

(A) Genetic makeup (B) Actual physical appearance
(C) Recessive alleles (D) None of the above

23. The idea that different pairs of alleles are passed to offspring independently is Mendel's principle of

(A) Unit inheritance (B) Segregation
(C) Independent assortment (D) All of these

24. The idea that for any particular trait, the pair of alleles of each parent separate and only one allele from each parent passes to an offspring is Mendel's principle of

(A) Independent assortment (B) Hybridization
(C) Segregation (D) None of these

25. Which of the following is totally impossible outcome of Mendel's Experiment?

(A) 3 tall 1 short plant (B) 24 tall and 8 short plants
(C) 8 tall and 0 short plants
(D) 4 tall plants and 1 medium height plant

26. Which of the following is not a direct conclusion that can be drawn from Mendel's Experiment?

(A) Only one parental trait is expressed
(B) Two copies of each trait is inherited in sexually reproducing organism
(C) For recessive trait to be expressed, both copies should be identical
(D) Natural selection can alter frequency of an inherited trait.

27. Which one is a possible progeny in F2 generation of pure bred tall plant with round seed and short plant with wrinkled seeds?
(A) Tall plant with round seeds
(B) Tall plant with wrinkled seeds
(C) Short plant with round seed
(D) All of the above

28. Which section of DNA provides information for one protein
(A) Nucleus (B) Chromosomes
(C) Trait (D) Gene

29. Mode of pollination in maize is–
(A) Self-pollination (B) Vegetative propagation
(C) Cross-pollination (D) None

30. TZ test colour of living tissues of seed changed to–
(A) Red (B) Blue
(C) Yellow (D) Green

31. Maximum moisture content for safe storage in wheat seed is–
(A) 12 (B) 15
(C) 7 (D) None

32. The structure and function of nitrogenase can be studied by–
(A) NMR (B) EPR
(C) Massbauer spectroscopy (D) All

33. In India, normally how many generation system seeds are produced?
(A) 1 (B) 2
(C) 3 (D) 4

34. Multiple effects of a single gene is known as
(A) Polyploidy (B) Heteroploidy
(C) Pleiotropy (D) None of these

35. Which system is used for hybrid seed production in onion?
(A) Dioecious (B) Cytoplasmic male sterility
(C) Genetic male sterility (D) Self incompatibility

36. What is seed multiplication ratio in potato and groundnut are respectively?
(A) 1:4 and 1:8 (B) 1:8 and 1:4
(C) 1:8 and 1:10 (D) 1:10 and 1:8

37. A mating among the following can not lead to heterosis
(A) AAbb × aaBB
(B) AABBccdd × AABBccdd
(C) AABBccdd × aabbCCDD
(D) aaBBcc × AABBCC

38. Pure line variety is a
(A) Homozygous – Heterogenous
(B) Homozygous – Hemogenous
(C) Heterogenous – Heterogenous
(D) Heterogenous – Homozygous

39. Centre of diversity refers to the area where cultivated plant species and or their wild relatives show
(A) Low competition with unrelated species
(B) Highest variation
(C) High ecosystem diversity
(D) Highest variation and high ecosystem diversity

40. Production of seedless grapes require–
(A) Gibberellins (B) NAA
(C) Ethylene (D) IAA

41. Inflorescence of cauliflower is called–
(A) Catkin (B) Head
(C) Curd (D) None

42. Pusa snow ball is variety of–
(A) Sapota (B) Cauliflower
(C) Ber (D) Cabbage

43. Development of embryo without fertillization is–
(A) Apomixis (B) Amghimixis
(C) Parthenocarpy (D) None

44. Which of the following is not part of a seed?
(A) Stored food (B) The cotyledon
(C) The fruit (D) The embryo

45. The Parthenogenesis refers to
(A) With seeds after pollination
(B) Without seeds after pollination
(C) With viable seeds after fertilization
(D) Seedless without fertilization

46. An example of naturally occurring parthenocarpic fruit
(A) Guava (B) Mango
(C) Banana (D) Apple

47. In which of the following plants, cotyledons form the first pair of leaves?
(A) Maize (B) Castor
(C) Rice (D) Bean

48. Seedless fruits in vitis are formed due to
(A) Sterility (B) Double fertilization
(C) Syngamy (D) Parthenocarpy

49. Which blood type would not be possible for children of a type AB mother and a type A father?
(A) O (B) A
(C) B (D) AB

50. Long radishes crossed with round radishes result in all oval radishes. This type of inheritance is:

(A) Multiple alleles (B) Complete dominance
(C) Co-dominance (D) Incomplete dominance

51. If two white sheep produce a black offspring, the parent's genotypes for colour must be:

(A) Heterozygous
(B) Homozygous white
(C) Homozygous black
(D) Not enough information was given

52. An extra finger in humans is rare but is due to a dominant gene. When one parent is normal and the other parent has an extra finger but is heterozygous for the trait, what is the probability that the first child will be normal?

(A) 0% (B) 25%.
(C) 50% (D) 75%

53. Nucellar polyembryony is reported in species of

(A) Brassica (B) Citrus
(C) Gossypium (D) Triticum

54. Endosperm of angiosperms results after fertilization from

(A) Synergids (B) Secondary nucleus
(C) Antipodal cells (D) Egg

55. Fertilization in which male gametes are carried through pollen tube is known as

(A) Porogamy (B) Syngamy
(C) Siphonogamy (D) Chalazogamy

56. Fragrant flowers with well developed nectaries are adapted for

(A) Anemophily (B) Hydrophily
(C) Entomophily (D) Zoophily

57. Germ pore is the region where the exine is

(A) Thin (B) Uniform
(C) Think and Uniform (D) Absent

58. How many microspore mother cell will produce 1000 microspores?

(A) 100 (B) 200
(C) 250 (D) 500

59. Compound layering is also called as

(A) Air layering (B) Serpentine layering
(C) Trench layering (D) Mound layering

60. Which of the following method is suitable for combining the desirable characters of two plants together in a single plant ?

(A) Layering (B) Cutting
(C) Grafting (D) All of these

61. In grafting, the rooted plant is called the stock. The stem cutting from the donor plant is called
(A) Transplant (B) Scion
(C) Bud graft (D) Root stock

62. Roses can grow successfully if stock and scion are
(A) Of the related species (B) Of different species
(C) Of different groups (D) Of similar class

63. For successful grafting, stock and scion shall be
(A) Of different species (B) Of same species
(C) Of similar groups (D) Of different class

64. Cutting is not suitable for
(A) Lime (B) Sugarcane
(C) Oleander (D) Tapioca

65. Which one is the reproductive unit having an embryo, reserve food and protective covering
(A) Spore (B) Fruit
(C) Seed (D) Fruitlet

66. In exalbuminous seeds, there is
(A) Nonformation of endosperm
(B) Abundant formation of endosperm
(C) Abundant protein
(D) Consumption of endosperm during development of seed

67. In exalbuminous seed, the food is generally stored in
(A) Endosperm initially but hypocotyls later on
(B) Cotyledons
(C) Endosperm from the beginning
(D) Testa

68. The seed in which reserve food is present in the perisperm
(A) Nymphaea (B) Coconut
(C) Onion (D) Litchi

69. What are the characteristics of petals?
(A) Attached to the receptacle
(B) Known as the corolla
(C) Produces pollen
(D) Missing in bees pollinated plants

70. Most seeds will not germinate until they are in favorable environmental conditions including:
(A) Temperature (B) Water
(C) Oxygen (D) All the above

71. What are siliquas?
(A) Splits along two seams, but the seeds are attached to a central partition
(B) Splits along two seams
(C) Splits along one seam (D) All the above

72. Which of the following is the correct definition for style?
(A) Connects the stigma & ovary
(B) At the top of the pistil
(C) Swollen base
(D) All the above

73. A hyaline, bisexual and self-fertilized flower that does not open at all, is
(A) Chasmogamous (B) Apogamous
(C) Cleistogamous (D) Polygamous

74. Axillary buds are used to raise crop of
(A) Wheat (B) Rice
(C) Groundnut (D) Sugarcane

75. Banana plant develops from
(A) Rhizome (B) Seed
(C) Sucker (D) Stolon

76. In a flowering plant, archesporium gives rise to
(A) Only the wall of the sporangium
(B) Both wall and the sporogenous cells
(C) Wall and the tapetum
(D) Only tapetum and sporogenous cells

77. A polygonum type of embryo sac is
(A) 7-celled and 8-nucleate (B) 8-celled and 7-nucleate
(C) 7-celled and 7-nucleate (D) 8-celled and 8-nucleate

78. Allogamy is best favoured by
(A) Dichogamy (B) Dicliny
(C) Cleistogamy (D) Homogamy

79. Dry dehiscent fruits can be extracted by
(A) Drying (B) Maceration
(C) Polishing (D) Threshing

80. Decortication is used for
(A) Groundnut (B) Maize
(C) Bhendi (D) Sunflower

81. Wet extraction is commonly used in
(A) Bittergourd (B) Rice
(C) Onion (D) Chillies

82. Pre cleaning mainly used to improve
(A) Vigour (B) Viability
(C) Physical purity (D) Genetic purity

83. Pre-cleaning of seeds is called
(A) Stratification (B) Scarification
(C) Scalping (D) Upgrading

84. The temperature limit for artificial air drying
(A) 100°C (B) 110°F
(C) 110°C (D) 65°F

85. Hard seed coat can be removed using
(A) Stratification (B) Debearders
(C) Aspirator (D) Pebble mill

86. Heated air drying limit is
(A) 110°F (B) 110°C
(C) 98°C (D) 98°F

87. If silica gel absorbs moisture, which colour it will develop?
(A) Blue (B) Pink
(C) Red (D) White

88. High moistened seed should be dried
(A) At high temperature (B) At low temperature
(C) Under direct sun (D) Low temperature shade dried

89. Oblong sieve is used for grading
(A) Paddy (B) Sorghum
(C) Maize (D) Sunflower

90. Flat and round seeds can be separated by
(A) Specific gravity separator (B) Spiral separator
(C) Magnetic separator (D) Indent cylinder separator

91. What is the top screen size for wheat seed processing
(A) 7/64" (B) 8/64"
(C) 12/64" (D) 14/64"

92. Commonly used fungicide for seed treatment is
(A) Furadan (B) Metasystox
(C) Carbendazim (D) Atrazine

93. Which one of the following chemical is used against bacteria?
(A) Carbendazim (B) Mancozeb
(C) Carboxyl (D) Streptomycin

94. Seed propagation is the rule in ________
(A) Cosmos (B) Cornflower
(C) Jasmine (D) Gypsophila

95. The area covered for field count is ________ in the case of line sowing medium spaced seed crops like chickpea
(A) 85% (B) 75%
(C) 60-70% (D) 60%

96. Match the following

List I	**List II**
(a) Capitulam	1. Coriander
(b) Umbel	2. Amaranthus
(c) Compound umbel	3. Sunflower
(d) Spike	4. Carrot

(A) 4 3 2 1 (B) 3 1 4 2
(C) 2 1 4 3 (D) 4 2 1 3

97. Match the following

List I	List II
(a) Calyx	1. Petals
(b) Corolla	2. Carpel
(c) Stigma	3. Androecium
(d) Anther	4. Sepal

(A) 4, 1, 2, 3 (B) 4, 2, 1, 3
(C) 2, 1, 4, 3 (D) 3, 2, 4, 1

98. Match the following

List I	List II
(a) Superior ovary	1. Epigynous
(b) Inferior ovary	2. Hypogynous
(c) Syncarpous	3. Multiple carpels
(d) Apocarpous	4. Fused carpels

(A) 4, 1, 2, 3 (B) 3, 2, 4, 1
(C) 2, 1, 4, 3 (D) 4, 2, 1, 3

99. Match the following

List I	List II
(a) Bisexual	1. Pisillate flower
(b) Unisexual	2. Perfect flower
(c) Monoecious	3. Date palm
(d) Dioecious	4. Arecanut

(A) 4, 1, 2, 3 (B) 2, 1, 4, 3
(C) 4, 2, 1, 3 (D) 3, 2, 4, 1

100. Match the following

List I	List II
(a) Andromonoecious	1.
(b) Gynomonoecious	2. + +
(c) Trimonoecious	3. +
(d) Hermaphrodite	4. +

(A) 4, 3, 2, 1 (B) 3, 2, 4, 1
(C) 4, 1, 2, 3 (D) 2, 3, 4, 1

Answer Keys

Q.No.	Ans.	Q.No.	Ans.	Q.No.	Ans.	Q.No.	Ans.	Q.No.	Ans.
1	B	21	A	41	C	61	B	81	A
2	A	22	B	42	B	62	A	82	C
3	C	23	C	43	C	63	B	83	C
4	A	24	C	44	C	64	A	84	D
5	A	25	D	45	D	65	C	85	B
6	D	26	D	46	C	66	D	86	D
7	C	27	D	47	B	67	B	87	B
8	C	28	D	48	D	68	A	88	C
9	D	29	C	49	A	69	B	89	A
10	B	30	A	50	D	70	D	90	B
11	C	31	A	51	A	71	A	91	B
12	D	32	D	52	C	72	D	92	C
13	A	33	C	53	B	73	C	93	D
14	B	34	C	54	B	74	D	94	D
15	A	35	B	55	C	75	C	95	C
16	D	36	A	56	C	76	B	96	B
17	A	37	B	57	D	77	A	97	A
18	C	38	B	58	C	78	A	98	C
19	B	39	D	59	B	79	D	99	B
20	A	40	A	60	C	80	A	100	A

MODEL TEST 15

1. **Days taken for physiological maturity of papaya seed is_________ days after anthesis**
   (A) 120-127 (B) 126-133
   (C) 132-139 (D) 138-145
2. **Who has suggested the factors that are responsible for genetic contamination in natural fields due to natural crossing?**
   (A) Kadam (1942) (B) Bateman (1947)
   (C) Dettann (1953) (D) Martin (1942)
3. **Which from the following crop is an example for emasculation and dusting method of hybrid seed production?**
   (A) Rice (B) Redgram
   (C) Maize (D) Cotton
4. **Pseudo-fertility may also occur late in the season as in**
   (A) Tomato (B) Tobacco
   (C) Tapioca (D) Tea
5. **In variety release, the best entries are proposed for which trial?**
   (A) RYT (B) PYT
   (C) MLT (D) ART
6. **Which one of the following is not relevant for land selection for seed production?**
   (A) Moderate water facilities (B) Good fertile soil
   (C) Neutral pH (D) Good drainage facilities
7. **Which is not relevant for over or excess irrigation of seed crops?**
   (A) Promote vegetative growth (B) Promote lodging
   (C) Promote nutrient imbalance (D) Delayed flowering
8. **Rainfall, temperature, light, wind and humidity are**
   (A) Climatic factors (B) Edaphic factors
   (C) Biotic factors (D) Topographic factors
9. **Which one of the following is not applicable for calculating the seed production requirement while seed planning?**
   (A) Seed replacement rate
   (B) Planting ratio
   (C) Seed rate of A, B and R lines
   (D) Seed yield of A, B and R lines

10. What is the safe moisture for cereal seed?

(A) 15% (B) 14%
(C) 12% (D) 18%

11. Where MSSRF is located?

(A) Chennai (B) Hyderabad
(C) Bangalore (D) Mumbai

12. Which one of the following is not a function of threshers?

(A) To grade the seed and deliver in the outlet.
(B) To clear the seed and deliver in the outlet.
(C) To thresh the seed out of the head
(D) To thresh the seed from the straw

13. What is the seed recovery percentage in HCl method of extraction of ashgourd?

(A) 4-5% (B) 5-6%
(C) 6-7% (D) 7-8%

14. For drying of paddy seeds, the suitable drying method is ___

(A) Non mixing column driers
(B) Rectangular metal bin driers
(C) Circular metal bin driers
(D) Square metal bin driers

15. Which from the following machine is engaged for upgrading the seed based on weight?

(A) Air screen cleaner cum grader
(B) Indented cylinder separator
(C) Specific gravity separator
(D) Magnetic separator

16. What is the suggested temperature during seed priming?

(A) 0-5°C (B) 5-10°C
(C) 10-15°C (D) 15-20°C

17. Which one of the following method is used for mixing and dividing of chaffy seeds?

(A) Spoon method (B) Random cup method
(C) Modified halving method (D) Hand halving method

18. Freedom from inert matter and defective seeds:

(A) Genetic purity (B) Physical purity
(C) Defective purity (D) Normal purity

19. Which one of the following is not correct for germinable seed while evaluating TZ test.

(A) Plumule unstained
(B) Entire embryo stained bright red
(C) Extremeties of scutellum unstained
(D) Extremeties of radicle tip unstained

20. Which one of the following method is not the direct method of seed moisture estimation?

(A) Toluene distillation method (B) Air oven method
(C) Moisture meter method (D) Lyophilization method

21. At what temperature and duration, the seeds are dried in oven using high constant temperature method for moisture determination?

(A) 110° ± 3°C for 4h (B) 110° ± 3°C for 6h
(C) 130° ± 2°C for 4h (D) 130° ± 3°C for 4h

22. In which container, the seed sample should be packed for testing moisture content at STL

(A) Moisture resistant container
(B) Moisture proof container
(C) Moisture vapour proof container
(D) Moisture pervious container

23. What is the size of working sample for purity analysis for okra

(A) 140 g (B) 70 g
(C) 120 g (D) 60 g

24. Which one of the following sample is used for purity analysis, germination test and other tests in the STL?

(A) Primary sample (B) Composite sample
(B) Submitted sample (D) Working sample

25. Which one of the following crop seed responds well for phenol test?

(A) Sorghum (B) Wheat
(C) Pearlmillet (D) Maize

26. Which one of the following seed has beak that is used for variety identification?

(A) Redgram (B) Castor
(C) Paddy (D) Groundnut

27. Where Royal Botanic Garden is located?

(A) USA (B) Germany
(C) London (D) Paris

28. Which one of the following is not stored under cryopreservation?

(A) Anther and pollen (B) Ovary and ovule
(C) Shoot apex (D) Meristem tip

29. Which one of the following is not related to new consensus of seed deterioration?

(A) Impaired transcription
(B) Production of free radicals causing incomplete enzyme synthesis damages to the cell

(C) Hindering of mitochondrial replication
(D) Lesser ATP synthesis

30. Which one of the following pest is a pre harvest infester?
(A) Wasps (B) Beetles
(C) Bruchids (D) Moths

31. In cryogenic storage, seeds are stored at
(A) 75°C (B) -192°C
(C) -100°C (D) -65°C

32. What is the dosage of carbendazim for seed treatment?
(A) 1-3 g/kg of seeds (B) 2-4 g/kg of seeds
(C) 3-5 g/kg of seeds (D) 4-6 g/kg of seeds

33. Eradication of fungal pathogen which are deep-seated inside the seed tissues is known as __________.
(A) Seed protection (B) Seed disinfection
(C) Seed disinfestation (D) Seed fumigation

34. In hydration and dehydration treatment, the seeds are soaked to raise the seed moisture upto __________.
(A) 15-20% (B) 20-25%
(C) 25-30% (D) 30-35%

35. Growing on test is adopted to test the presence of ______ .
(A) Fungi (B) Virus
(C) Nematode (D) Bacteria

36. Which one of the following is not a fruiting structures of fungi that can be observed in examination of dry seed?
(A) Acervuli (B) Mycelium
(C) Sclerotia (D) Pycnidia

37. Which one of the following common name for *Tribolium surinamens* is
(A) Khapra beetle (B) Red flour beetle
(C) Saw-tooth grain beetle (D) Confused flour beetle

38. Which fungi infects the seed through seed parts?
(A) *Fusarium oxysporum* (B) *Xanthomonas compestris*
(C) *Ustilaga tritici* (D) *Rhizoctonia solani*

39. Which pathogen contaminates episperm and that is systemic transmission?
(A) Sclerospora
(B) *Colletotrichum truncatum*
(C) *Neovossia indica*
(D) *Corynebacterium michiganense*

40. Which disease is the designated one of greengram?
(A) Black arm (B) Ashy stem blight
(C) Downy mildew (D) Red rot

41. Isolation distance for FS of cole crops like cabbage and cauliflower is

(A) 800 m (B) 200 m
(C) 400 m (D) 1600 m

42. Match the following

a. Scarification - 1. Tree crops
b. Hot water treatment - 2. Pulses
c. Leaching - 3. Coriander
d. Stratification - 4. Cole crops

(A) 2 1 3 4 (B) 3 4 1 2
(C) 4 3 1 2 (D) 2 4 3 1

43. Optimum seed moisture for storage of paddy is

(A) 7 - 8% (B) 10 - 12%
(C) > 13% (D) < 8%

44. In hybrid seed production, pollination for 100% seed set can be achieved by

(A) Supplementary pollination
(B) Rope pulling
(C) Rod driving
(D) All the above

45. Pure seed percentage for certified seed in wheat is about

(A) 98% (B) 95%
(C) 99.9% (D) 85%

46. As per ISTA standards, final count of seed germination in rice is taken on

(A) 7th day (B) 14th day
(C) 10th day (D) 3rd day

47. First Seed Testing Laboratory in India was established in

(A) 1969 (B) 1961
(C) 1983 (D) 1971

48. SMR for groundnut is

(A) 1:40 (B) 1:8
(C) 1:80 (D) 1:20

49. Maximum percentage of inert matter in maize as per minimum seed standard

(A) 0.5% (B) 0.2%
(C) 2% (D) 5%

50. Isolation distance for hybrid maize (FS) is about

(A) 800 m (B) 400 m
(C) 200 m (D) 10 m

51. Planting ratio of paddy in hybrid seed production is

(A) 10:2 (B) 4:2
(C) 2:1 (D) 12:8

52. Maximum permitted percentage of off type for CS production in sorghum is
(A) 0.5% (B) 1.0%
(C) 0.1% (D) None of these

53. Prechilling treatment required crops in seed germination test is
(A) Paddy (B) Cauliflower
(C) Pulses (D) Chillies

54. Resting stage of seed is a
(A) Quiescence (B) Endogenous dormancy
(C) Thermo dormancy (D) Photo dormancy

55. The crop in which dormancy is advantageous
(A) Sesame (B) Paddy
(C) Blackgram (D) Groundnut

56. Standard seed germination percentage for chillies is
(A) 45% (B) 75%
(C) 60% (D) 90%

57. Quantity of other crop seeds permitted for CS of bhendi is
(A) 5 Nos/kg (B) 10 Nos/kg
(C) 15 Nos/kg (D) 25 Nos/kg

58. Isolation distance for FS of tomato is
(A) 400 m (B) 600 m
(C) 200 m (D) 50 m

59. Time isolation is followed for
(A) Sorghum (B) Rice
(C) Pearl millet (D) Cotton

60. Optimum temperature for seed priming is
(A) 0 - 5°C (B) 5 - 10°C
(C) 10 - 15°C (D) 15 - 20°C

61. Emasculation in hybrid cotton field is done in
(A) Male line (B) Female line
(C) Sterile line (D) Both (A) and (B)

62. Pure seed percentage of groundnut is
(A) 99% (B) 98%
(C) 97% (D) 96%

63. Heliotropism is a special character of
(A) Paddy (B) Groundnut
(C) Sunflower (D) Sugarcane

64. Maximum permitted offtype in castor for FS is
(A) 1.0% (B) 0.5%
(C) 2.0% (D) 5.0%

65. Quicker method of seed extraction in tomato is
(A) Alkali method (B) Acid method
(C) Fermentation method (D) Mechanical method

66. Number of field inspections in chilli is
(A) 3 (B) 4
(C) 1 (D) 2

67. Tea is an example of _____ seed.
(A) Recalcitrant (B) Orthodox
(C) Difficult recalcitrant (D) Both (A) and (B)

68. TIBA is an/a
(A) Growth regulator (B) Growth retardant
(C) Gametocide (D) Growth promotor

69. Maximum percentage of designated disease in all cole crops (FS) is
(A) 0.50% (B) 0.10%
(C) 0.2% (D) 0.05%

70. Match the following

a. CMS - 1. Cauliflower
b. GMS - 2. Rice
c. CGMS - 3. Redgram
d. Self incompatibility - 4. Vegetative propagated crops

(A) 2 4 1 3 (B) 2 3 4 1
(C) 4 3 2 1 (D) 1 3 2 4

71. The quality of gametocide which is not relevant is
(A) Its effect should be uniform in whole blooming period
(B) Minimum side effect on plant growth
(C) Its effectiveness should not be influenced by environment
(D) It is contact in nature

72. Minimum germination percentage is low for
(A) Bitter gourd (B) Paddy
(C) Maize (D) Redgram

73. Brinjal is a
(A) Cross pollinated crop (B) Often cross pollinated crop
(C) Self pollinated crop (D) None of these

74. Identify the wrongly matched option relevant to moisture content in storage under moisture pervious container
(A) Pumpkin - 7.0%
(B) Capsicum - 8.0%
(C) Cotton - 6%
(D) Maize - 12%

75. Maximum permitted shedding tassel percentage in hybrid maize production is
(A) 2% (B) 1%
(C) 0.1% (D) 0.5%

76. Pearl millet is cross pollinated due to its
(A) Protandrous nature (B) Protogynous nature
(C) Hermaphrodite nature (D) All the above

77. The isolation distance is higher in
(A) Self pollinated crop (B) Cross pollinated crop
(C) Often cross pollinated crop (D) Both (A) and (B).

78. Hardening of Pearl millet seed is done with
(A) 2% KH_2PO_4 (B) 2% KCl
(C) Soaking in water (D) 100 ppm $ZnSO_4$

79. Male sterility used in rice hybrid seed production is
(A) Cytoplasmic male sterility
(B) Genetic male sterility
(C) Cytoplasmic genetic male sterility
(D) Both (A) and (B)

80. Seed drying is very important to maintain its
(A) Protein content (B) Oil content
(C) Viability and vigour (D) Chemical composition

81. Seed which is commercially used for crop production by the farmers is
(A) Foundation seed (B) Certified seed
(C) Breeder seed (D) Nucleus seed

82. In hybrid seed production, the seed rate of male line is ____ female line.
(A) Higher than (B) Lower than
(C) Equal to (D) Depends on crop

83. Planting ratio of female to male in cotton hybrid seed production is
(A) 4:2 (B) 8:2
(C) 2:1 (D) 20:2

84. Identify the incorrect statement
(i) 2, 4-D spraying in brinjal at flowering stage increase fruit set.
(ii) Minimum germination percentage of tomato is 70%
(iii) Isolation distance for bhendi FS is 400 m
(iv) Seed rate in TPS production is higher compared seed tuber production in potato

(A) (iv) alone is incorrect (B) (ii) and (iii) are incorrect
(C) All of these (D) (i) alone is incorrect

85. Match the following

	Operation	**Crop**
a.	Failing and pounding	- 1. Cereals
b.	Stripping	- 2. Rice
c.	Animal treading	- 3. Pulses
d.	Power tiller threshing	- 4. Groundnut

(A) 1, 4, 2, 3 (B) 2, 4, 1, 3
(C) 4, 3, 1, 2 (D) 3, 4, 1, 2

86. Find out the incorrect statement.

(i) Once over harvest - Bhendi
(ii) 6 - 8 harvests - Paddy
(iii) Once over harvest - Paddy

(A) (iii) alone incorrect
(B) (i) alone is incorrect
(C) (i) and (ii) are incorrect
(D) All are incorrect

87. Find out the correct statement

(i) Seed recovery percentage of acid method is 1.0%
(ii) Seed recovery percentage of fermentation method is 1.0%
(iii) Seed recovery percentage of alkali method is 1.5%
(iv) The chemical used in acid method is NaCl

(A) (iii) and (iv) are correct (B) (i) alone is correct
(C) (iii) alone is correct (D) (i) and (ii) are correct

88. LSU dryer is a

(A) Batch type dryer (B) Continuous flow dryer
(C) Column dryer (D) Rectangular bin dryer

89. Match the following

a. Seed priming - 1. Taylor
b. Solid matrix priming - 2. Callon
c. Matri conditioning - 3. Khan
d. Bio priming - 4. Malnassy

(A) 3, 4, 2, 1 (B) 1, 3, 4, 2
(C) 4, 1, 3, 2 (D) 2, 3, 4, 1

90. Find out the correct statement

(i) Seed fortification involves hydration and dehydration
(ii) Seed hardening involves hydration and dehydration

(A) (i) alone is correct (B) (ii) alone is correct
(C) Both are correct (D) Both are incorrect

91. Which of the following is adhesive in seed pelleting?

(A) Lime
(B) Rock phosphate
(C) *Albizia amara* leaf powder
(D) Gelatin

92. ____________ is a filler used in seed pelleting

(A) Gum arabic (B) Gelatin
(C) Gypsum (D) Methyl cellulose

93. Find out the incorrect statement

(A) Pelleted seed can't be stored for more than 3 to 4 months
(B) Seed polymer coating is suitable for all type of lands
(C) Active principle in beetroot for seed colouring is betanin
(D) Seed colouring results in breakage

94. Which of the following statement is related to the *ex-situ* conservation?

(A) Established protected area network, with appropriate management practices, corridors to link fragments, restore degraded habitats within and outside.

(B) Established botanical and zoological gardens, conservation stands, bank of germplasm for pollen, seed, seedlings, tissue culture, gene and DNA etc.

(C) Agriculture is totally dependent on ecosystem processes and functions provided by biodiversity, soil formation, nutrient cycling and pollination of crops, maintaining water cycles

(D) None of the above

95. Adventive embryony in citrus is due to

(A) Nucellus
(B) Integuments
(C) Zygotic embryo
(D) Fertilised egg

96. Match the following

	Vegetables		Spacing
(a)	Chillies	1.	75×60 cm
(b)	Brinjal	2.	45×45 cm
(c)	Tomato	3.	60×30 cm
(d)	Bhendi	4.	75×75 cm

(A) 4, 2, 3, 1
(B) 2, 1, 4, 3
(C) 3, 1, 4, 2
(D) 4, 3, 2, 1

97. Match the following

	Crop		Maximum inert matter
(a)	Cluster bean	1.	1
(b)	Bhendi	2.	4
(c)	Sorghum	3.	2
(d)	Groundnut	4.	3

(A) 3, 2, 1, 4 (B) 4, 3, 2, 1
(C) 1, 2, 4, 3 (D) 3, 1, 4, 2

98. Match the following

	Crop		Working sample (g)
(a)	Niger	1.	16
(b)	Mustard	2.	75
(c)	Kidney bean	3.	120
(d)	Triticale	4.	15

(A) 4, 2, 1, 3 (B) 4, 1, 2, 3
(C) 1, 2, 4, 3 (D) 2, 1, 3, 4

99. Match the following

	Crop		**Working sample (gram)**
(a)	Fenugreek	1.	90
(b)	Cotton Hybrid	2.	4
(c)	Indian clover	3.	25
(d)	Daincha	4.	10

(A) 2, 3, 4, 1 (B) 4, 3, 2, 1
(C) 4, 2, 1, 3 (D) 2, 3, 4, 1

100. Match the following

(a)	Root rot	1.	Verticillium
(b)	Wilt	2.	Pythium
(c)	Internally seed borne	3.	Seed surface
(d)	Externally seed borne	4.	Endosperm

(A) 2, 3, 4, 1 (B) 2, 4, 3, 1
(C) 2, 1, 4, 3 (D) 1, 3, 4, 2

Answer Keys

Q.No.	Ans.	Q.No.	Ans.	Q.No.	Ans.	Q.No.	Ans.	Q.No.	Ans.
1	B	**21**	D	**41**	D	**61**	B	**81**	B
2	B	**22**	C	**42**	A	**62**	C	**82**	B
3	D	**23**	A	**43**	B	**63**	C	**83**	B
4	B	**24**	D	**44**	D	**64**	B	**84**	A
5	D	**25**	B	**45**	A	**65**	B	**85**	B
6	A	**26**	D	**46**	B	**66**	A	**86**	C
7	D	**27**	C	**47**	B	**67**	A	**87**	D
8	A	**28**	B	**48**	B	**68**	C	**88**	B
9	A	**29**	B	**49**	C	**69**	B	**89**	C
10	C	**30**	C	**50**	B	**70**	C	**90**	B
11	A	**31**	B	**51**	A	**71**	D	**91**	D
12	A	**32**	B	**52**	C	**72**	A	**92**	C
13	B	**33**	B	**53**	B	**73**	C	**93**	D
14	A	**34**	A	**54**	A	**74**	C	**94**	B
15	C	**35**	D	**55**	D	**75**	B	**95**	A
16	C	**36**	B	**56**	C	**76**	B	**96**	B
17	D	**37**	D	**57**	A	**77**	B	**97**	D
18	D	**38**	B	**58**	D	**78**	B	**98**	B
19	A	**39**	A	**59**	B	**79**	C	**99**	A
20	C	**40**	B	**60**	C	**80**	C	**100**	C

MODEL TEST 16

1. **Development of fruits without fertilization is called**
   (A) Apospory (B) Parthenogenesis
   (C) Parthenocarpy (D) Polygamy

2. **The water of coconut is**
   (A) Liquid mesocarp (B) Liquid endocarp
   (C) Liquid endosperm (D) Liquid nucellus

3. **The fruit of rice and wheat is called**
   (A) Achene (B) Siliqua
   (C) Follicle (D) Caryopsis

4. **If any somatic cell of sporophyte produces gametophyte without reduction division, it is**
   (A) Parthenogenesis (B) Apogamy
   (C) Apospory (D) Amphimixis

5. **Development of a fruit without fertilization is known as**
   (A) Parthenogenesis (B) Polyembryo
   (C) Budding (D) Apomixes

6. **Development of seed without sexual process is**
   (A) Apomixes (B) Polyembryony
   (C) Parthenogenesis (D) Polygamous

7. **Which one of the following organ is a food storage tissue of peanut?**
   (A) Megagametophyte (B) Endosperm
   (C) Cotyledons (D) Perisperm

8. **Starch is a polymer of**
   (A) Fructose (B) Glucose
   (C) Dextrose (D) Galactose

9. **Amylose is a straight chain polymer of glucose units joined by __________ linkages**
   (A) Alpha 1-6 glycosidic (B) Beta 1-6 glycosidic
   (C) Alpha 1-4 glycosidic (D) Beta 1-4 glycosidic

10. **Which one of the following storage protein is soluble in salt solution?**
    (A) Albumin (B) Globulin
    (C) Prolamin (D) Glutelin

11. **An organic substance that can withstand environmental extremes cannot be degraded by any enzyme is**
(A) Lignin (B) Cellulose
(C) Cuticle (D) Sporopollenin

12. **Caryopsis type of fruit is present in which crop?**
(A) Groundnut (B) Corn
(C) Blackgram (D) Greengram

13. **A winged achene is**
(A) Nut (B) Achene
(C) Samara (D) Siliqua

14. **Which one of the following is a schizocarp**
(A) Raspberry (B) Strawberry
(C) Pineapple (D) Carrot

15. **Dry indehiscent single-seeded fruit formed from bicarpellary syncarpous inferior ovary is:**
(A) Cremocarp (B) Caryopsis
(C) Cypsela (D) Berry

16. **Flowers are zygomorphic in**
(A) Datura (B) Mustard
(C) Gulmohur (D) Tomato

17. **A head or capitulum is characteristic of**
(A) Compositae (B) Umbelliferae
(C) Solanaceae (D) Graminae

18. **Which one of the following protects the plumule?**
(A) Epicotyl (B) Hypocotyl
(C) Coleoptile (D) Coleorhiza

19. **Miniature plant contains one or two seed leaves called**
(A) Radicle (B) Plumule
(C) Cotyledon (D) Zygote

20. **Which of the following is a non-endospermic seed**
(A) Cereals (B) Millets
(C) Orchids (D) Custard apple

21. **The life cycle of an angiosperm is consist of**
(A) Sporophytic stage (B) Ovule stage
(C) Ovary stages (D) Placentation stage

22. **The suspensor formed in angiosperm seed development links**
(A) Roots to shoots (B) Stem to leaves
(C) Embryo to nutrients (D) Cotyledons to epicotyl

23. **If any somatic cell of sporophyte produces gametophyte without reduction division, it is _____**
(A) Parthenogenesis (B) Apogamy
(C) Apospory (D) Amphimixis

24. Somatic embryoids are
(A) identical with zygotic embryos and without seed coats
(B) identical with zygotic embryos and with seed coats
(C) non-identical with zygotic embryos and without seed coats
(D) non-identical with zygotic embryos and with seed coats

25. In which one pair both the plants can be vegetatively propagated by leaf pieces?
(A) Asparagus and Bryophyllum (B) Agave and Kalanchoe
(C) Bryophyllum and Kalanchoe (D) Chrysanthemum and Agave

26. Which of the following statement is true regarding vegetative propagation?
(A) Vegetative propagation occurs naturally and artificially producing genetically identical plants
(B) Vegetative propagation occurs naturally and artificially producing genetically different plants
(C) Vegetative propagation occurs artificially producing genetically similar plants
(D) Vegetative propagation occurs naturally producing genetically similar plants

27. In asparagus one or more short, green cylindrical or sometimes flattened branches of limited growth develop from node of stem, such a branch is known as
(A) Cladode (B) Phylloclade
(C) Stem tendril (D) Thorn

28. When the other floral parts are arranged at the base of the gynoecium, the flower is called
(A) Perigynous flower (B) Epigynous flower
(C) Hypogynous flower (D) None of these

29. When the ovule is bent round like a horse shoe it is said to be
(A) Orthotropous (B) Anatropous
(C) Amphilotropous (D) Campylotropous

30. When the pistil has all the carpels united together it is said to be
(A) Apocarpous (B) Syncarpous
(C) Simpler types (D) None of the above

31. Which part of a seed contains the inhibitory substance
(A) Cotyledon (B) Embryo
(C) Scutellum (D) Pericarp

32. Among the following which compound can induce seed dormancy?
(A) Abscisic acid (B) Potassium nitrate
(C) Gibberellins (D) Ethylene

33. Mention the number of days taken to reach physiological maturity in cotton

(A) 50 days after anthesis with boll cracking
(B) 52 days after anthesis with boll cracking
(C) 54 days after anthesis with boll cracking
(D) 56 days after anthesis with boll cracking

34. Which of the following is not true about synthetic seeds?

(A) Cab be stored for a year without the loss of valuables
(B) Easy to handle
(C) Can be directly sown in the soil like natural seeds
(D) Need hardening in the greenhouse

35. Genetic self-incompatibility tends to increase

(A) Pollen dispersal (B) Inbreeding
(C) Pollination (D) Isolation

36. When the pollen unites with the ovule to form a seed it is called

(A) Pollination (B) Fertilization
(C) Germination (D) Anthesis

37. Wind pollinated plants differ from insect pollinated plants in having

(A) Small petals and sticky pollen
(B) Small coloured petals and heavy pollen
(C) Coloured petal and large pollens
(D) No petals and light pollen

38. Fragrant flowers with well developed nectaries are adapted for

(A) Anemophily (B) Hydrophily
(C) Entomophily (D) Zoophily

39. The condition in which the anthers and stigmas of a bisexual flower mature at the same time is known as

(A) Cleistogamy (B) Heterogamy
(C) Syngamy (D) Homogamy

40. Certification is not required for :

(A) Foundation seed I (B) Breeder seed
(C) Foundation seed II (D) Certified seed

41. In which one of the following country, DLF-Trifolium is located?

(A) Denmark (B) Switzerland
(C) Germany (D) Japan

42. Which of the following seed fertication treatment is adopted for paddy.

(A) 1% $CaCl_2$ (B) 2% KCl
(C) 0.5% Potassium dihydrogen chloride
(D) 100 ppm $ZnSO_4$

43. Which one of following organic solvent is used for seed infusion?
(A) Ketone (B) Acetone
(C) Acetyl CoA (D) Acetic acid

44. What is the suggested temperature during seed priming?
(A) 0-5°C (B) 5-10°C
(C) 10-15°C (D) 15-20°C

45. Which one of the following is not relevant for seedling evaluation?
(A) Normal seedlings (B) Abnormal seedlings
(C) Hard seeds (D) Fresh seeds

46. The number of healthy normal seedlings / number of seed tested × 100 =
(A) Percent viability (B) Percent germination
(C) Percent purity (D) Percent healthy seed

47. Which one of the following is the maximum permissible off types for 99.7% genetic purity?
(A) 0.10% (B) 0.20%
(C) 0.30% (D) 0.50%

48. What colour tag is used for International seed lot certificate?
(A) White (B) Green
(C) Blue (D) Orange

49. Which type of vigour test is GADA test?
(A) Physical test (B) Performance test
(C) Biochemical test (D) Stress test

50. Who has worked out the vigour index?
(A) Hiltner (B) Moore
(C) Abdul-Baki and Anderson (D) Mathews and Bradnock

51. Difference between seed and grain is
(A) Seed quality (B) Seed processing
(C) Genetic purity (D) Price

52. Registered seed tag colour is
(A) Azure blue (B) White
(C) Golden yellow (D) Purple

53. Test weight of seeds is called
(A) 1000 seeds (B) 100 seeds
(C) 50 seeds (D) 10 seeds

54. Self pollinated crop is
(A) Groundnut (B) Brinjal
(C) Pigeon pea (D) (A) & (C)

55. Safe seed storage moisture content is
(A) 12-15% (B) 10-12%
(C) 12-14% (D) 7-8%

56. Swelling of seed due to imbibition of water by

(A) Water potential (B) Turgor pressure
(C) Osmosis (D) None

57. Seed viability testing method

(A) Tetrazolium chloride test (B) Embryo culture test
(C) Petridish method (D) (A) & (B)

58. Isolation distance for sunflower certified seed production is

(A) 500 m (B) 100 m
(C) 50 m (D) 10 m

59. Often cross pollinated crops are

(A) Cotton (B) Brinjal
(C) Pigeon pea (D) (A) & (C)

60. National seed project

(A) 1978 (B) 1968
(C) 1988 (D) 1958

61. Phases of seed certification is

(A) 5 (B) 6
(C) 7 (D) 8

62. Slow drying seed is

(A) Grass (B) Pulses
(C) Cereals (D) Rapeseed and mustard

63. Tetrazolium test is used to determine the level of

(A) Dehydrogenase (B) ATP ase
(C) Carboxylase (D) Isomerase

64. Alkaloid test is used for

(A) Cotton (B) Brinjal
(C) Pigeon pea (D) Lupins

65. Hot water treatment is used to break dormancy in

(A) Brinjal (B) Cherry
(C) Lentil (D) Cotton

66. Germination is inhibited by

(A) IR light (B) UV light
(C) Red light (D) Green light

67. Seed drying is very important to maintain its

(A) Oil content (B) Viability
(C) Protein content (D) Nutrient content

68. Safe storage moisture content of wheat is

(A) 12-15% (B) 10%
(C) 12% (D) 7%

69. New genes are created by

(A) Mutation (B) None
(C) Inversion (D) Recombination

70. Semi root parasite is
(A) Sorghum (B) Lucerne
(C) Striga (D) Loranthus

71. Seed is defined as
(A) Matured ovule (B) Embryo enclosed in seed coat
(C) Any propagation material (D) All

72. National organisation involved in seed production is
(A) ISST (B) TDC
(C) TUCAS (D) NSC

73. Foundation seed is the progeny of
(A) TFL seed (B) Foundation seed I
(C) Nucleus seed (D) Breeder seed

74. Isolation distance for foundation seed crop of tomato is
(A) 50 m (B) 100 m
(C) 400 m (D) 600 m

75. Best seed extraction method for tomato is
(A) Acid method (B) Alkali method
(C) Manual method (D) Machine extraction

76. Hybridisation technique followed for tomato is
(A) Use of chemicals (B) Use of CMS lines
(C) Emasculation & Dusting (D) Self incompatibility

77. Protogynous flower occur in
(A) Tomato (B) Brinjal
(C) Chilli (D) None

78. For hybrid seed production of bhendi, male and female lines are raised in
(A) Use of chemicals (B) Use of CMS lines
(C) Emasculation & Dusting (D) Block method

79. True to type seed and healthy seed of onion are produced by
(A) Seed to seed method (B) Bulb to seed method
(C) Both (D) None of the above

80. A cauliflower fruit is called
(A) Siliqua (B) Drupe
(C) Berry (D) Caryopsis

81. Optimum seed moisture for bitter gourd seed storage ?
(A) 8% (B) 12%
(C) 6% (D) 14.5%

82. What is the method used for wet seed extraction in gourds ?
(A) Cutting and scarping
(B) Acid method
(C) Manual scooping and washing with water
(D) None of the above

83. Pollination behaviour of cowpea is
(A) Self (B) Cross
(C) Often cross (D) None

84. One of the following which doesn't cross with cauliflower
(A) Radish (B) Mustard
(C) Chinese cabbage (D) All

85. Ideal temperature for bolting in cabbage is
(A) 10-15°C (B) 4-7°C
(C) 3-5°C (D) None

86. Over wintering is essential for flowering in
(A) Bhendi (B) Chilli
(C) Chow-chow (D) Cauliflower

87. Best suited season for planting TPS
(A) April- May (B) May – June
(C) June – July (D) July – August

88. The temperature limit for heated air drying is
(A) 110ºF (B) 65ºF
(C) 87ºF (D) 168ºF

89. Which is orthodox type of seed?
(A) Mango (B) Jack
(C) Citrus (D) Tomato

90. Combines are used for
(A) sowing of seeds (B) harvesting the crops
(C) threshing (D) harvesting and threshing

91. Separating seeds from chaff is called
(A) winnowing (B) threshing
(C) fallow (D) harvesting

92. Seed drill is used to
(A) sow the seeds (B) remove the weeds
(C) remove the pest (D) mix manure in the soil

93. Calcium chloride absorbs moisture ______ % of its weight
(A) 10 (B) 15
(C) 20 (D) 25

94. During processing, care should be taken to avoid
(A) Mechanical damage (B) Heat damage
(C) Frost damage (D) Threshing injury

95. Specific gravity separator upgrade the seeds according to their
(A) Weight (B) Length
(C) Surface texture (D) Colour

96. Match the following

(a) *Xanthomonas*	1. Systemic infection through flowers
(b) *Pseudomonas*	2. Systemic infection through stigma
(C) *Ustilago*	3. Systemic infection of seed
(D) *Alternaria*	4. Stick on the surface of the seed

(A) 4, 3, 2, 1 (B) 3, 4, 1, 2
(C) 3, 2, 1, 4 (D) 3, 1, 2, 4

97. Match the following

(a) Paddy	1. Smut
(b) Wheat	2. Kernel smut
(c) Pearl millet	3. Bunt
(d) Sorghum	4. Downy mildew

(A) 1, 3, 2, 4 (B) 4, 2, 3, 1
(C) 3, 1, 4, 2 (D) 3, 1, 2, 4

98. Match the following

(A) Rice weevil	1. *Rhizopertha dominica*
(B) Lesser grain borer	2. *Sitophilus oryzae*
(C) Red flour beetle	3. *Trichoderma granaria*
(D) Khapra beetle	4. *Tribolium castanium*

(A) 2, 1, 4, 3 (B) 3, 1, 4, 2
(C) 3, 2, 4, 1 (D) 4, 1, 2, 3

99. Match the following

(A) Maize	1. Red flour beetle
(B) Wheat	2. Angoumois moth
(C) Rice	3. Khapra beetle
(D) Pulses	4. *Callosobruchus chinensis*

(A) 4, 2, 3, 1 (B) 1, 3, 2, 4
(C) 1, 2, 3, 4 (D) 3, 4, 2, 1

100. Match the following

(A) Rice	1. Cyst nematode
(B) Rye	2. White tip nematode
(C) Tomato	3. Ear cockle nematode
(D) Pigeon pea	4. Root knot nematode

(A) 1, 3, 2, 4 (B) 3, 1, 4, 2
(C) 2, 3, 4, 1 (D) 1, 2, 3, 4

Answer Keys

Q.No.	Ans.	Q.No.	Ans.	Q.No.	Ans.	Q.No.	Ans.	Q.No.	Ans.
1	C	21	A	41	A	61	B	81	C
2	C	22	C	42	C	62	B	82	C
3	D	23	C	43	B	63	A	83	A
4	C	24	A	44	C	64	D	84	A
5	A	25	C	45	D	65	C	85	A
6	A	26	A	46	B	66	A	86	D
7	C	27	A	47	C	67	B	87	A
8	B	28	C	48	D	68	C	88	A
9	C	29	C	49	C	69	D	89	D
10	B	30	B	50	C	70	C	90	B
11	D	31	D	51	C	71	D	91	A
12	B	32	A	52	D	72	D	92	A
13	C	33	B	53	A	73	D	93	A
14	D	34	B	54	A	74	A	94	A
15	C	35	C	55	C	75	A	95	A
16	C	36	B	56	C	76	C	96	D
17	A	37	D	57	D	77	C	97	C
18	C	38	C	58	A	78	D	98	A
19	C	39	D	59	D	79	B	99	B
20	C	40	B	60	C	80	A	100	C

MODEL TEST 17

1. **Generation model used in seed production of cross pollinated crops**
(A) Three generation model (B) Two generation model
(C) Five generation model (D) All

2. **Seed act was enforced on**
(A) 2nd oct 1969 (B) 2nd oct 1968
(C) 2nd oct 1966 (D) 2nd oct 1972

3. **National seed research and training centre is located at**
(A) Lucknow (B) Dehradun
(C) Bhopal (D) Varanasi

4. **Tag colour of truthfully labeled seeds is**
(A) Golden yellow (B) Azure blue
(C) purple (D) Opal green

5. **Which of the following is having epigeal germination?**
(A) Rajma (B) Chick pea
(C) Pea (D) Groundnut

6. **Phenol and NAOH tests are recommended for which of the following?**
(A) Rice (B) Wheat
(C) Maize (D) Sorghum

7. **Concept of foundation seed production was given by**
(A) Paul neelgard (B) J.W. Robertson
(C) N.A. Nobbe (D) S.V. Marthur

8. **Tag size of breeder seed is**
(A) 12 × 6 cm (B) 15 × 7.5 cm
(C) 15 × 10 cm (D) 12 × 7.5 cm

9. **Sugar formed in germinating seed is**
(A) Dextrose (B) Maltose
(C) Sucrose (D) Germinose

10. **Mites don't survive below**
(A) 10°C (B) 5°C
(C) 9°C (D) 0°C

11. **According to Harrington thumb rule, for every decrease of ______ °C in storage temperature, seed life gets doubled.**
(A) 15°C (B) 10°C
(C) 5°C D.25°C

12. Peroxidase test is used for

(A) Soybean (B) Wheat
(C) Cotton (D) Sorghum

13. Designated disease of potato is

(A) Root knot (B) Scab
(C) Wart of potato (D) All the above

14. Male to female ratio in hybrid seed production of sorghum is

(A) 2:4 (B) 2:6
(C) 2:8 (D) 3:1

15. Purity percent is lowest in which of the following

(A) Carrot (B) Pearl millet
(C) Rape seed and mustard (D) Sesame

16. Germination percentage is lowest in which of the following?

(A) Cotton (B) Wheat
(C) Carrot (D) Cabbage

17. Designated weed of berseem is

(A) *Convolvulus arvensis* (B) *Citrullus colosynthes*
(C) Senji (Melilotus) (D) *Chicorium intybus*

18. Isolation distance for wheat foundation seed production

(A) 10 m (B) 20 m
(C) 5 m (D) 3 m

19. Seed multiplication ratio of potato is

(A) 1:8 (B) 1:10
(C) 1:4 (D) 1:15

20. Non diffusible red colour compound formed in tetrazolium test is

(A) Formazone (B) Ketazone
(C) Feluric acid (D) Caumarins

21. The International Crop Improvement Association (renamed as ASCA) was formed in

(A) 1919 (B) 1920
(C) 1923 (D) 1925

22. In 1946, the classes of quality seeds was given for the first time for

(A) Pulse crops (B) Forage crops
(C) Vegetable crops (D) Grain crops

23. Modern classification of improved seed recognizes the following _______ broad classes.

(A) 4 Classes (B) 3 Classes
(C) 5 Classes (D) 6 Classes

24. Breeder seed can be produced from

(A) Nucleus seed (B) Breeder seed
(C) Both (A) & (B) (D) Foundation seed

25. Minimum field inspection recommended in case of potato is
(A) 3 (B) 4
(C) 5 (D) 2

26. Designated disease of wheat is
(A) Loose smut (B) Head smut
(C) Alternaria blight (D) Leaf roll

27. Impurity percentage of seed is called as
(A) Impurity (B) Blockage
(C) Dockage (D) Inert

28. KOH Bleach test is used for
(A) Wheat (B) Maize
(C) Barley (D) Sorghum

29. Chemical used in case of tetrazolium test is
(A) 2,3,4-triphenyl tetrazolium chloride
(B) 2,3,5-triphenyl tetrazolium chloride
(C) 2,3,5-triphenyl tetrazolium bromide
(D) 2,4,5-triphenyl tetrazolium bromide

30. In oven method of moisture content determination, seed samples are dried at _____ °C temperature for 90 mins.
(A) 140°C (B) 150°C
(C) 105°C (D) 130°C

31. National seed corporation was initiated on
(A) 7th March, 1963 (B) 5th March, 1963
(C) 3rd oct, 1963 (D) 2nd oct, 1963

32. Minimum isolation distance for foundation seed production of Brinjal is
(A) 400 m (B) 250 m
(C) 300 m (D) 150 m

33. Pure seed percentage for groundnut is
(A) 97 (B) 98
(C) 96 (D) 99

34. Permissible amount of total weed seeds ______ (kg) for radish seed foundation seed
(A) 10 (B) 5
(C) 15 (D) 20

35. Moisture content for hybrid maize is
(A) 10% (B) 12%
(C) 9% (D) 15%

36. Which of the following is not available in India ?
(A) Certified seed (B) TFL seed
(C) Registered seed (D) Foundation seed

37. **Which of the following is an example of negative photoblastic seeds?**
(A) Tobacco (B) Ficus
(C) Tomato (D) Rhododendron

38. **A variety that has to be protected is called as**
(A) Extant variety (B) Candidate variety
(C) Breeders variety (D) Protected variety

39. **Which of the following is an example of endospermic seed ?**
(A) Castor (B) Mango
(C) sunflower (D) Tamarind

40. **SRR for self pollinated crops is**
(A) 25% (B) 45%
(C) 55% (D) 65%

41. **Hot water treatment is used to break dormancy in case of**
(A) *Brassica campestris* (B) *Prunus cerasus*
(C) Lentil (D) *Havea* sp.

42. **Tetrazolium test was given by**
(A) Lakon (B) Nobbe
(C) Robertson (D) Neergard

43. **Indian seed act was first time amended at**
(A) 1972 (B) 1975
(C) 2000 (D) 2001

44. **The sample formed by combining and mixing all the primary sample is**
(A) Primary sample (B) Composite sample
(C) Submitted sample (D) Working sample

45. **First seed production programme for temperate vegetables started in India at**
(A) Hyderabad (B) Punjab
(C) Akola (D) Quetta

46. **Germination percentage of __________ is highest in case of field crops**
(A) Cotton (B) Maize
(C) Wheat (D) Pearl millet

47. **SMR of paddy is**
(A) 1:80 (B) 1:40
(C) 1:20 (D) 1:10

48. **Toxic substance found in case of sorghum**
(A) HCN (B) CN glucoside
(C) Erucic acid (D) Saponin

49. **Carrot is grown as __________ crop for seed purpose.**
(A) Annual (B) Biennial
(C) Perennial (D) None of the above

50. Weight of one bale of jute is
(A) 160 kg (B) 170 kg
(C) 180 kg (D) 190 kg

51. Which of the following is not a basic requirement for protection of variety under PBR ?
(A) Novelty (B) Distinctness
(C) Uniformity (D) High yield

52. Ear to row method is related to
(A) Nuclear seed (B) Foundation seed
(C) Registered seed (D) Certified seed

53. Objectionable weed of wheat is
(A) *Argemone mexicana* (B) *Convulvulus arvensis*
(C) *Abelmoschus* sp. (D) Wild cucurbits

54. The pollination behaviour of palm tree
(A) Self pollinated (B) Cross pollinated
(C) Often cross pollinated (D) Self and cross pollinated

55. Maintainer line in hybrid seed production plot is called as
(A) R line (B) B line
(C) Restorer line (D) Both (A) & (C)

56. Block system is used for the hybrid seed production in
(A) Cotton (B) Maize
(C) Sorghum (D) Castor

57. Synchronized flowering between parental lines of bajra is achieved by
(A) Detasseling (B) Jerking
(C) Rope pulling (D) GA Spray

58. Staggered sowing is practiced for
(A) Good field experiment (B) Tolerate to pest and disease
(C) Uniform flowering (D) Easy processing

59. GA is sprayed to hybrid seed production plots for
(A) Delayed flowering (B) Enhanced flowering
(C) To induce flowering (D) Complete exertion of flowering

60. Isolation distance could be modified based on the number of border rows in
(A) Maize (B) Sorghum
(C) Pearl millet (D) Castor

61. The chemical used for dormancy breaking in sunflower is
(A) Gibberellic acid (B) Kinetin
(C) Ethrel (D) Coumarin

62. The volunteer plants are
(A) weeds
(B) Diseased plants
(C) Low vigour Plants
(D) Plants from left over seeds of last season

63. The objectionable weed of watermelon is
(A) *Cucumis hardwickii* (B) *Cucumis sativus*
(C) *Cucumis lanatus* (D) *Cucumis colocynthis*

64. Ganga 5 is a hybrid variety of
(A) Maize (B) Sorghum
(C) Bajra (D) Rice

65. The best method for seed extraction in tomato is
(A) Acid method (B) Alkali method
(C) Fermentation method (D) All

66. In a seed production, mechanical admixtures in a seed lot affects the
(A) Physical purity (B) Genetic purity
(C) Keeping quality of seed (D) Physiological quality

67. The breeding tool for cotton hybrid seed production
(A) CMS (B) Emasculation and dusting
(C) GMS (D) Self incompatibility

68. Intermediate between auto and allogamous
(A) Maize (B) Cotton
(C) Rice (D) Tomato

69. An albuminous dicot seed is
(A) Castor (B) Peas
(C) Beans (D) Mango

70. In synthetic seed, the following part is the natural
(A) Endosperm (B) Embryo
(C) Seed coat (D) None

71. The isolation distance recommended to avoid contamination with Johnson grass in hybrid sorghum is
(A) 200 m (B) 100 m
(C) 400 m (D) 50 m

72. Round and flat seed in a seed lot can be separated by
(A) Draper belt or velvet roll mill
(B) Indent cylinder seperator
(C) Specific gravity separator
(D) Debearder

73. Weed seed, other crop seed and inert matter are fractions of
(A) Purity analysis (B) Seed viability test
(C) Germination (D) All of the above

74. The standard for germination (minimum) for certified seed of gram is
(A) 85% (B) 90%
(C) 80% (D) 70%

75. TZ test colour of living tissues of seed changed to
(A) Red (B) Blue
(C) Yellow (D) Green

76. Loose smut is
(A) Internally seed borne (B) Externally seed borne
(C) Air borne (D) None

77. Solar treatment is used for
(A) Loose smut (B) Stem rust
(C) Powdery mildew (D) Blast

78. Seedlessness in fruits is called as
(A) Parthenocarpy (B) Parthenogenesis
(C) Apomixis (D) None

79. Male sterility is used to develop
(A) Inbreds (B) Synthetics
(C) Hybrids (D) None

80. Half gram is working sample for purity analysis of
(A) Tomato (B) Onion
(C) Berseem (D) Tobacco

81. While sampling from a seed lot of 50 bags, the minimum number of primary sample should be
(A) 5 (B) 10
(C) 15 (D) 20

82. Production of seedless grapes requires
(A) Gibberellins (B) NAA
(C) Ethylene (D) IAA

83. Epigeal germination is found in
(A) Pea (B) Gram
(C) Cucumber (D) Mango

84. Reproduction by leaf occurs in
(A) Begonia (B) Hydrilla
(C) Ceratophyllum (D) None

85. Basic input of agriculture is
(A) Seed (B) Fertilizer
(C) Soil (D) Irrigation

86. Dicot endospermic seed is
(A) Castor (B) Fenugreek
(C) Both (A) & (B) (D) Bean

87. The chemical seed treatment is given to control
(A) Fungi & bacteria (B) Soil insects
(C) Storage insects (D) All

88. Which one of the following seed requires coarse grinding before drying for moisture determination?
(A) Bajra (B) Sorghum
(C) Soybean (D) Ragi

89. Based on which, the moisture content of seeds is determined

(A) Electrical conductivity (B) Moisture conductivity
(C) Physical conductivity (D) Mechanical conductivity

90. Clipping off a portion of seed coat is done to break dormancy of which kind of seed?

(A) Large seed (B) Small seed
(C) Minute seed (D) Tiny seed

91. What may be the reason for non germination of seeds due to immature embryo?

(A) Endogenous dormancy (B) Exogenous dormancy
(C) Primary dormancy (D) Secondary dormancy

92. What is the suitable temperature and relative humidity for accelerated ageing test?

(A) 40±2°C and 95±2% (B) 40±2°C and 90±2%
(C) 41±2°C and 95±2% (D) 41±2°C and 90±2%

93. There are few approaches to varietal identification which are furnished below. Find out the incorrect one

(A) Pathological approaches (B) Physiological approaches
(C) Biochemical approaches (D) Molecular approaches

94. FORM-I for seed certification should be submitted within ____________ from the date of transplanting

(A) 5 days (B) 10 days
(C) 15 days (D) 20 days

95. The production of high quality and uniform embryos has been limited to only.

(A) Carrot (B) Alfalfa
(C) Both (A) and (B) (D) Sandalwood

96. Match the following

Crop		Seed rate
(a) Rice	1.	10 kg/ha
(b) Maize	2.	7.5 – 10 kg/ha
(c) Sorghum	3.	3.75 kg/ha
(d) Pearl millet	4.	60 kg/ha

(A) 4, 1, 2, 3 (B) 1, 2, 3, 4
(C) 2, 3, 4, 1 (D) 3, 2, 4, 1

97. Match the following

Crop		Seed rate
(a) Blackgram	1.	15-20 kg/ha
(b) Cotton	2.	20-25 kg/ha
(c) Sesamum	3.	10-15 kg/ha
(d) Safflower	4.	2 – 4 kg/ha

(A) 3, 4, 2, 1 (B) 4, 3, 2, 1
(C) 2, 1, 4, 3 (D) 2, 1, 4, 3

98. Match the following

Crop	Seed rate
(a) Pumpkin	1. 1.5 kg/ha
(b) Ashgourd	2. 5 - 7 kg/ha
(c) Bittergourd	3. 2.5 kg/ha
(d) Snakegourd	4. 6 - 8 kg/ha

(A) 4, 3, 2, 1 (B) 2, 1, 4, 3
(C) 4, 1, 2, 3 (D) 3, 1, 4, 2

99. Match the following

Hybrid crop	Seed rate
(a) Pearl milllet	1. 4 to 4.5 ♀ & 1 to 1.5 ♂ kg/ha
(c) Sunflower	2. 6 ♀ & 4 ♂ kg/ha
(c) Cotton	3. 3.5 ♀ & 0.75 ♂ kg/ha
(d) Castor	4. 8 to 10 ♀ & 3 to 3.5 ♂ kg/ha

(A) 4, 3, 1, 2 (B) 4, 2, 3, 1
(C) 3, 1, 4, 2 (D) 2, 3, 4, 1

100. Match the following

Hybrid crop	Isolation distance
(a) Sorghum	1. 100 m
(b) Rice	2. 400 m
(c) Chilli	3. 30 m
(d) Cotton	4. 200 m

(A) 4, 3, 2, 1 (B) 2, 4, 3, 1
(C) 4, 1, 3, 2 (D) 4, 1, 2, 3

Answer Keys

Q.No.	Ans.	Q.No.	Ans.	Q.No.	Ans.	Q.No.	Ans.	Q.No.	Ans.
1	A	21	A	41	C	61	??	81	B
2	A	22	B	42	A	62	D	82	A
3	D	23	B	43	A	63	D	83	C
4	D	24	C	44	B	64	A	84	A
5	A	25	A	45	D	65	A	85	A
6	B	26	A	46	B	66	A	86	C
7	B	27	C	47	A	67	B	87	D
8	A	28	D	48	A	68	B	88	C
9	B	29	B	49	B	69	A	89	A
10	B	30	D	50	C	70	B	90	A
11	C	31	A	51	D	71	??	91	A
12	A	32	C	52	A	72	A	92	B
13	D	33	C	53	B	73	A	93	A
14	A	34	B	54	B	74	A	94	C
15	A	35	B	55	B	75	A	95	A
16	C	36	C	56	A	76	A	96	A
17	D	37	C	57	B	77	A	97	C
18	D	38	B	58	C	78	A	98	A
19	C	39	A	59	D	79	C	99	C
20	A	40	A	60	A	80	C	100	D

MODEL TEST 18

1. 32nd triennial International Seed Testing Association (ISTA. Congress 2019 was held at

(A) Bangalore (B) Mumbai

(C) Hydrabad (D) Varanasi

2. When 32nd triennial International Seed Testing Association (ISTA) Congress 2019 was held

(A) 26 June – 3 July, 2019 (B) 25 June – 2 July, 2019

(C) 6 June – 13 June, 2019 (D) 16 June – 23 June, 2019

3. The National Bureau of Agriculturally Important Insects was established by ICAR in _______, was earlier known as _______

(A) Bangalore; PDBC

(B) New Delhi; National Pusa Collection

(C) Ranchi; Indian Lac Research Institute

(D) New Delhi; NCIPM

4. The term 'Heterosis' was coined by

(A) G.H. Shull (B) W. Bateson

(C) T.H. Morgan (D) E.M. East

5. Who found that a peculiar regularity exists in the ratios of nitrogenous bases of double stranded DNA?

(A) James Watson and Francis Crick

(B) Rosalind Franklin

(C) Erwin Chargaff

(D) Waclaw Szybalski

6. Which one of the following classes of plant hormones is used as a substitute for low temperature (pre-chilling) seed treatment?

(A) Cytokinin (B) Gibberellins

(C) Ethylene (D) Auxins

7. "Hot water" seed treatment for management of bacterial leaf blight of rice is done at

(A) 43°C for 20 minutes (B) 53°C for 30 minutes

(C) 59°C for 20 minutes (D) 62°C for 30 minutes

8. The entire process of aerobic respiration occurs in

(A) Cytosol (B) Mitochondria

(C) Mitochondria and cytosol (D) Mitochondria and peroxisomes

9. In germinating oil seeds, beta-oxidation machinery operates in

(A) Oleosomes (B) Glyoxysomes
(C) Lysosomes (D) Mitochondria

10. The first naturally occurring cytokinin was isolated from

(A) DNA of herring sperm (B) Endosperm of coconut milk
(C) Immature corn seeds (D) DNA of human's liver

11. Who isolated first naturally occurring cytokinin

(A) Miller (B) Lakon
(C) Hull (D) Harrington

12. Which of the following concepts was first introduced by Toshio Murashige?

(A) Gene Libraries (B) Single Cell Protein (SCP)
(C) Synthetic Seed (D) C-DNA

13. Photomorphogenic responses are most evident in germinating seedlings. The most effective wavelength for this phenomenon is

(A) 560 nm (B) 660 nm
(C) 680 nm (D) 730 nm

14. The growth of pollen tube in the pistil is an example of positive chemotropism. This chemotropic movement is in response to

(A) Potassium gradient (B) Calcium gradient
(C) Magnesium gradient (D) Phosphorus gradient

15. In wheat, maize and rice

(A) Perisperm is fused with seed coat
(B) Pericarp is fused with seed coat
(C) Perisperm and seed coat are separate
(D) Pericarp and seed coat are separate

16. Perisperm is present in the seeds of

(A) Coffee (B) Castor
(C) Pumpkin (D) Wheat

17. "Florigen" that induces flowering is made up of which two classes of hormones?

(A) Gibberellin and cytokinin (B) Gibberellin and anthesin
(C) Cytokinin and abscisic acid (D) Cytokinin and anthesis

18. Filiform apparatus is found in the

(A) Egg cell (B) Suspensor
(C) Endosperm (D) Synergids

19. Dormancy caused by seed coat impermeability to water and gases can be broken by

(A) Scarification (B) Stratification
(C) Low temperature (D) Abscisic acid

20. Perisperm in the seed is a
(A) Persistent endosperm (B) Persistent nucellus
(C) Persistent hypostase (D) Persistent epistase

21. Plant breeders are interested in apomixis because it
(A) Enlarges genetic variability
(B) Improves seed set
(C) Yields clonal progeny
(D) Shortens the life cycle

22. Which one of the following pathogens infect unfertilized ovary in pearlmillet?
(A) *Claviceps purpurea* (B) *Claviceps fusiformis*
(C) *Claviceps paspali* (D) *Claviceps Africana*

23. Which of the following dyes is used in testing the viability of protoplasts?
(A) Evans blue (B) Trypan blue
(C) Fluroscein diacetate (D) Brilliant blue

24. Which among the following defines a situation where the **Which among the following defines a situation where the** **pollen grains from the anther of a flower is transferred to the stigma of another flower on the same plant?**
(A) Xenogamy (B) Geitonogamy
(C) Autogamy (D) Herkogamy

25. Epigeal germination occurs in
(A) Castor (B) Garden pea
(C) Corn (D) Gram

26. Ethylene is used to overcome seed dormancy. Which of the following is used as precursor in synthesis of ethylene?
(A) Threonine (B) Aspartic acid
(C) Lysine (D) Methionine

27. Which of the following inorganic substances has a great effect on pollen germination and pollen tube growth?
(A) Copper (B) Zinc
(C) Boron (D) Molybdenum

28. Seed lot certificate colour is
(A) Orange (B) Green
(C) Purple (D) Orange or Green

29. Hot water treatment is used to break dormancy in
(A) Lentil (B) Cherry
(C) Cotton (D) None

30. Mak – 12 is a variety of
(A) Maize (B) Medicago
(C) B*t* Cotton (D) Tomato

31. Difference between seed and grain is
(A) Seed processing (B) Genetic purity
(C) Seed quality (D) Cost

32. Fanning mill is
(A) Air screen cleaner (B) Hammer mill
(C) Blower (D) Rice mill

33. National Nematode collection of India is located at
(A) New Delhi (B) Dharwad
(C) Anand (D) Bangalore

34. Seed coat peroxidase test is used for
(A) Sorghum (B) Wheat
(C) Soybean (D) Groundnut

35. The blotter method of seed health testing detects primarily
(A) Virus (B) Fungi
(C) Bacteria (D) Nematodes

36. A sound seed certification programme requires
(A) Direct participation of breeder
(B) Use of sophesticated equipment
(C) Classification of seed
(D) Support of law

37. The Rudimentary root of the seed or seedling that forms the primary root of the young plant is known as
(A) Rachis (B) Radicle
(C) Rachilla (D) Raceme

38. Mode of pollination in maize is
(A) Self-pollination (B) Vegetative propagation
(C) Cross-pollination (D) None

39. TZ test colour of living tissues of seed changed to
(A) Red (B) Blue
(C) Yellow (D) Green

40. Maximum moisture content for safe storage in wheat seed is
(A) 12 (B) 15
(C) 7 (D) None

41. The structure and function of nitrogenase can be studied by
(A) NMR (B) EPR
(C) Massbauer spectroscopy (D) All

42. In India, normally how many generation system seeds are produced?
(A) 1 (B) 2
(C) 3 (D) 4

43. Dicot endospermic seed is
(A) Castor (B) Fenugreek
(C) Both (A) and (B) (D) Bean

44. Endosperm nucleus of flower is developed into the
(A) Seed cell (B) Endosperm embryo
(C) Endosperm nucleus (D) Endosperm tissue

45. Endosperm nucleus is
(A) Haploid (B) Diploid
(C) Triploid (D) Tetraploid

46. After production of embryo and endosperm tissue, ovule turns into
(A) Stems (B) Root
(C) Seed (D) Fruit

47. Endosperm is rich in starch and
(A) Proteins (B) Lipids
(C) Carbohydrates (D) Sugar

48. After production of embryo and endosperm tissue, ovary turns into
(A) Seed (B) Fruit
(C) Stems (D) Roots

49. A seed is a
(A) Ripened ovule (B) Ripened ovary
(C) Ripened endosperm (D) Fused carpel

50. All the following structures are present in dicot seed except
(A) radical (B) endosperm
(C) coleoptiles (D) seed coat

51. Fruits are mature _______ whereas seeds are mature _____
(A) ovaries and stamens (B) ovules and carpels
(C) ovules and ovaries (D) ovaries and ovules

52. The seed coat is formed from
(A) integuments (B) ovary
(C) endosperm (D) pericarp

53. A seed contains
(A) stored food (B) embryo
(C) seed coat (D) all of these

54. Cotyledons are formed from
(A) embryos (B) seeds
(C) ovaries (D) ovules

55. All are dicotyledonous seeds except
(A) mango (B) castor
(C) rice (D) pea

56. Which of the following is a non-endospermic seed
(A) cereals (B) millets
(C) orchids (D) custard apple

57. The endosperm formed as a part of double fertilization is
(A) haploid (B) diploid
(C) triploid (D) polyploidy

58. Seed dormancy may be due to
(A) immature embryo
(B) hard seed coat
(C) presence of germination inhibitors
(D) all of these

59. Endosperm in gymnosperm is
(A) Haploid (B) Diploid
(C) Triploid (D) None

60. Endosperm in Angiosperms is
(A) Haploid (B) Diploid
(C) Triploid (D) None

61. Ovule is absent in
(A) Pteridophytes (B) Gymnosperm
(C) Angiosperm (D) All

62. Egg is present in
(A) Ovule (B) Venter
(C) Neck (D) None

63. The outer most layer of the sporogenous tissue forms the
(A) Tapetum (B) Spores
(C) Neck (D) Wall

64. Megasporangium is
(A) Pollen sac (B) Ovule
(C) Seed (D) Venter

65. Self pollination is a form of
(A) Inbreeding (B) Out breeding
(C) Random mating (D) None of these

66. Cross pollination is associated with
(A) Cleistogamy (B) Chasmogamy
(C) Dichogamy (D) Apogamy

67. Lower temperature than normal, convert male flowers into perfect flowers in ______
(A) Date palm (B) Mango
(C) Papaya (D) Citrus

68. Tomato is a ______ crop
(A) Self pollinated (B) Insect pollinated
(C) Often cross pollinated (D) Wind pollinated

69. Turmeric is propagated by
(A) Corms (B) Seed
(C) Rhizome (D) Bulb

70. Sweet potato is propagated by
(A) Stem cuttings (B) Tubers
(C) Roots (D) (A) & (B)

71. Hormones which control apical dominance are
(A) Auxins (B) Cytokinins
(C) Gibberellins (D) Polyamines

72. Tetraploid watermelon must be pollinated by __________ plants to produce a seedless watermelon
(A) Triploid (B) Aneuploid
(C) Diploid (D) Tetraploid

73. Water melon for seed should be harvested when
(A) Tendrils wither (B) Skin color is green
(C) Skin color is pale yellow (D) Vines dry out

74. During storage of vegetable seeds, seed moisture should be
(A) Equal to 10% (B) Greater than 10%
(C) Less than 10% (D) None of above

75. Removal of off type plants from a seed field is termed as
(A) Weeding (B) Rouging
(C) Hoeing (D) All of above

76. The most important seed component affecting seed /viability during storage is
A. Temperature (B) Humidity
(C) Moisture (D) Food reserves

77. Potato is generally propagated through
(A) Seed potato (B) Potato seed
(C) True Potato Seed (D) All of above

78. Isolation during seed production is done by
(A) Types (Varieties) (B) Plantation at a distance
(C) Zoning (D) All of above

79. Seed production in biennial vegetables is mostly done by
(A) Seed to seed method (B) In situ method
(C) Replanting method (D) All of above

80. Male sterility is not due to
(A) Genetic factor (B) Genome factor
(C) Cytogenetic factor (D) Nutritional factor

81. Hybrid vigor in vegetables is manifested by
(A) High yield (B) Uniformity in size
(C) Uniformity in maturity (D) All of above

82. Cross pollination is essential in following type of vegetables
(A) Monoecious (B) Dioecious
(C) Gynoecious (D) Perigynous

83. Cross pollination in plants bearing bisexual flowers is due to
(A) Dichogamy (B) Herkogamy
(C) Heterostyly (D) All of above

84. In anemophilous plants, pollens are disseminated by means of

(A) Insects (B) Water

(C) Wind (D) Gravity

85. In pollen development Gametocytes induces a problem of

(A) Sterility (B) Viability

(C) Fertility (D) Mobility

86. Nature of pollination in cucurbits is

(A) Highly cross pollinated (B) Oftenly cross pollinated

(C) Highly self pollinated (D) None of above

87. For hybrid seed production ______ flowers are completely suppressed on female plants.

(A) Staminate (B) Pistillate

(C) Staminode (D) Alternate

88. Controlled inhibition of seed is termed as

(A) Priming (B) Soaking

(C) Chilling (D) None of above

89. Celery and lettuce at high temperature become

(A) Photo dormant (B) Thermo dormant

(C) Ecodormant (D) Non dormant

90. Dominance of angiosperms can be considered due to evolution of

(A) Seed (B) Fruit

(C) Vascular system (D) Extensive root system

91. Fruit is a

(A) Post fertilization product of pistil

(B) Product of flower

(C) Body having seeds

(D) Product of ovary

92. A true fruit is formed from

(A) Ovary (B) Ovary and thalamus

(C) Ovary and calyx (D) Ovary and receptacle

93. A true fruit is

(A) Developed ovule

(B) Developed ovary

(C) Fertilized and developed ovary

(D) Fertilized and developed ovule

94. An accessory fruit is the one which develops from

(A) An ovary (B) Ovary and thalamus

(C) Unfertilized ovary (D) Inflorescence

95. Major function of phosphorus is in the formation of

(A) Cell membranes (B) Cell wall

(C) Enzymes (D) Carbohydrates

96. Match the following

	Season		**Crop**
(a)	December - January	1.	Sorghum hybrid
(b)	October - December	2.	Rice hybrid
(c)	November - December	3.	Cowpea variety
(d)	February - March	4.	Pearl millet hybrid

(A) 2, 3, 1, 4 (B) 2, 1, 4, 3
(C) 1, 2, 4, 3 (D) 2, 3, 4, 1

97. Match the following

	Crop		**Season**
(a)	Jute	1.	July and January
(b)	Blackgram	2.	November - December
(c)	Bhendi	3.	June - September
(d)	Ashgourd	4.	February - March

(A) 4, 3, 2, 1 (B) 4, 2, 1, 3
(C) 3, 4, 1, 2 (D) 3, 2, 4, 1

98. Match the following

	Age of seedling		**Crop**
(a)	25 - 30 days	1.	Brinjal
(b)	18 days	2.	Chilli
(c)	30 - 35 days	3.	Rice
(d)	28 - 42 days	4.	Pearl millet

(A) 4, 2, 1, 3 (B) 2, 4, 3, 1
(C) 3, 4, 1, 2 (D) 2, 1, 4, 3

99. Match the following

	Crop		**Foliar spray**
(a)	Field lab lab	1.	NAA 50 ppm
(b)	Cotton	2.	2% DAP
(c)	Tomato	3.	Ethephon 250 ppm
(d)	Ashgourd	4.	1% DAP

(A) 2, 4, 1, 3 (B) 2, 4, 3, 1
(C) 1, 3, 4, 2 (D) 1, 4, 3, 2

100. Match the following

	Hybrid crop		**Planting ratio**
(a)	Rice	1.	4:2
(b)	Pearl millet	2.	10:1
(c)	Brinjal	3.	6:1
(D)	Redgram	4.	8:2

(A) 4, 3, 1, 2 (B) 4, 2, 3, 1
(C) 3, 1, 4, 2 (D) 2, 3, 4, 1

Answer Keys

Q.No.	Ans.	Q.No.	Ans.	Q.No.	Ans.	Q.No.	Ans.	Q.No.	Ans.
1	C	21	C	41	D	61	D	81	D
2	A	22	B	42	C	62	A	82	B
3	A	23	C	43	C	63	A	83	D
4	A	24	C	44	D	64	B	84	A
5	A	25	A	45	C	65	A	85	A
6	B	26	D	46	C	66	B	86	A
7	B	27	C	47	A	67	C	87	A
8	B	28	D	48	B	68	A	88	D
9	D	29	A	49	A	69	C	89	B
10	C	30	C	50	C	70	A	90	B
11	A	31	B	51	D	71	B	91	D
12	C	32	A	52	A	72	C	92	A
13	B	33	A	53	D	73	A	93	B
14	B	34	C	54	A	74	A	94	B
15	B	35	B	55	C	75	B	95	A
16	B	36	A	56	C	76	A	96	B
17	B	37	B	57	C	77	D	97	A
18	D	38	C	58	D	78	B	98	C
19	A	39	A	59	A	79	D	99	A
20	B	40	A	60	C	80	D	100	A

MODEL TEST 19

1. **Raised nursery beds are _______ cm high from the ground level**

   (A) 5–10 (B) 15–20
   (C) 25–30 (D) 30–35

2. **A space of ___________ cm is left between two nursery beds**

   (A) 30–40 (B) 20–25
   (C) 15–20 (D) 20–25

3. **The nursery soil is sterilised in an autoclave at _________ °c for 30 minutes**

   (A) 100 (B) 140
   (C) 121 (D) 80

4. **In hot water seed treatment, seeds are placed at a temperature of ___________ °C for 10–30 minutes**

   (A) 48–55 (B) 30–35
   (C) 20–25 (D) 15–20

5. **Fungus used for seed treatment to control seed and soil-borne inoculums is ___________**

   (A) Trichogramma (B) Trichoderma
   (C) Rhizopus (D) Rhizobium

6. **The capability of a seed to germinate and produce normal seedlings is called ___________**

   (A) Seed vitality (B) Seed viability
   (C) Seed dormancy (D) Compatibility

7. **Seeds can be defined as a dormant __________**

   (A) Embryo (B) Ovule
   (C) Ovary (D) Fruit

8. **Which of the following is not a part of a seed?**

   (A) Seed coat (B) Embryo
   (C) Cotyledon (D) Ovary

9. **The process of loosening and turning the soil is called**

   (A) Broadcasting (B) Irrigation
   (C) Ploughing (D) Levelling

10. **The process of separating grain from chaff is called**

    (A) Threshing (B) Weeding
    (C) Sowing (D) Winnowing

11. Maintenance and multiplication of A line used in crops like pear millet, is done through
(A) Crossing the plants of 'A' line with those of 'B' line
(B) Crossing the plants of 'A' line with those 'R' line
(C) Selfing the plants of 'A' line
(D) Crossing the plants of 'A' line with any popular variety

12. Which system is used for hybrid seed production in onion?
(A) Self-compatibility
(B) Dioecious
(C) Cytoplasmic male sterility
(D) Genetic male sterility

13. The male gametic nuclei are the result of mitotic division for which one of the following?
(A) Male gametophyte (B) Microspores
(C) Microspore mother cell (D) Generative nucleus

14. Cross-pollination in Maize is mainly on account of which one of the following?
(A) Dioecism (B) Monoecism
(C) Protogyny (D) Self-compatibility

15. Which one of the following is the simplest method for testing the General Combining Ability of an inbred line?
(A) Single cross (B) Multiple cross
(C) Top cross (D) Test cross

16. Consider the following statements:
1. Acid lime (*Citrus pennivesiculata*) is propagated commercially through seeds.
2. Sweet lime (*Citrus limettioides*) is generally propagated by layering or hardwood cutting.

Which of the stataements given above is/are correct?
(A) 1 only (B) 2 only
(C) Both 1 and 2 (D) Neither 1 nor 2

17. In modern agriculture, which one of the following vector is commonly used for gene transfer from one species to another?
(A) Azotobacter (B) Agrobacterium
(C) E-coli (D) Neurospora

18. Isogenic lines are
(A) Lines identical in phenotype but having different genotypes
(B) Homozygous lines developed through continuous inbreeding
(C) Genotypically and phenotypically identical lines
(D) Lines identical in genotype except for one gene

19. Consider the following factors
1. Seed germination 2. Seed-borne diseases
3. Genetic purity 4. Weed seeds

The correct sequence in terms of importance (most important first, least important last) of these factor from the point of view of seed certification is

(A) 3, 2, 1, 4 (B) 2, 3, 1, 4
(C) 2, 3, 4, 1 (D) 3, 2, 4, 1

20. Synthetic seeds are
(A) Artificially synthesized seeds
(B) Somatic embryos encapsulated in suitable matrix
(C) Seeds of plants modified genetically
(D) None of these

21. Somatic embryoids are
(A) Identical with zygotic embryos and without seed coats
(B) Identical with zygotic embryos and with seed coats
(C) Non-identical with zygotic embryos and without seed coats
(D) Non-identical with zygotic embryos and with seed coats

22. The preserved embryoids are termed as
(A) Synthetic seeds
(B) Semi-synthetic seeds
(C) Natural seeds
(D) Fermented seeds

23. Encapsulation is necessary to produce and protect synthetic seeds. The encapsulation is carried out by various types of hydrogels, which are
(A) Soluble in water
(B) Soluble in organic solvents
(C) Insoluble in water
(D) Insoluble in organic solvents

24. Which of the following is not true about synthetic seeds?
(A) Can be stored for a year without the loss of valuables
(B) Easy to handle
(C) Can be directly sown in the soil like natural seeds
(D) Need hardening in the green house

25. Recalcitrant seeds are
(A) Resistant to drying and freezing temperature
(B) Killed by drying and freezing temperature
(C) Both (A) and (B)
(D) None of above

26. The encapsulation of somatic embryos can be carried out by
(A) Automatic encapsulation process
(B) Gel complexation
(C) Both (A) and (B)
(D) Coating protein

27. What is correct order of process and procedure of certification of seeds?

1. Seed sampling and analysis, including genetic purity test and/or seed health test, if any, in order to verify conformity to the prescribed standards
2. Grant of certificate and certification tags, tagging and sealing.
3. Receipt and scrutiny of application, verification of seed source, class and other requirements of the seed used for raising the seed crop.
4. Field inspections to verify conformity to the prescribed field standards.

(A) 3, 4, 1, 2 b. 1, 2, 3, 4
(C) 3, 4, 2, 1 d. 4, 3, 2, 1

28. Biotechnology is a tool that has been widely accepted worldwide for improvement of cultivars with regard to

(A) Insect pest resistance
(B) Tolerance to drought, cold and salinity condition
(C) Nutrition enhancement
(D) Post harvest quantity

Find out the incorrect answer

(A) 1 (B) 2
(C) 3 (D) 4

29. Match the following and choose the correct option.

a. Leaves (i) Anti-transpirant
b. Seed (ii) Transpiration
c. Roots (iii) Negative osmotic potential
d. Aspirin (iv) Imbibition
e. Plasmolyzed cell (v) Absorption

(A) a-(ii), b-(iv), c-(v), d-(i), e-(iii)
(B) a-(iii), b-(ii), c-(iv), d-(i), e-(v)
(C) a-(i), b-(ii), c-(iii), d-(iv), e-(v)
(D) a-(v), b-(iv), c-(iii), d-(ii), e-(i)

30. Match the following and choose the correct option.

a. Pollen grain 1. attracts pollinator.
b. Microsporangium 2. male gametophyte
c. Flower 3. enhances dispersal of seeds
d. Carpel 4. sterile jacket of cells

(A) 2, 4, 1, 2 (B) 2, 1, 3, 4
(C) 4, 3, 2, 1 (D) 1, 3, 4, 2

31. Plants produce seeds

(A) To feed animals. (B) To travel back in time
(C) To reproduce (D) To shade other plants

32. All seeds germinate
(A) Right next to the parent plant.
(B) Whenever the conditions are right for that type of plant
(C) Only after they've been eaten.
(D) As soon as they get a little wet.

33. What is the age of the oldest living seed found?
(A) 120 years (B) 3000 years
(C) 1200 years (D) 10,000 years

34. A plant embryo
(A) Always floats in the ocean.
(B) Is the new baby plant.
(C) Something outside the seed that gives the seed a clue about where it is.
(D) Is the thing that attaches a seed to it's mother.

35. A dormant seed is
(A) Developing but not growing
(B) Dead
(C) Growing
(D) Not growing or developing

36. Seeds respond to clues
(A) From their surroundings (B) From their mother plant
(C) From outer space (D) From other seeds

37. Xenia and metaxenia terms are related with
(A) Pollen culture (B) Only endosperm
(C) Xylem and Phloem (D) Pollen and endosperm

38. Xenia is effect of pollen grains on
(A) Embryo (B) Endosperm
(C) Embryo sac (D) Nucellus

39. In a seed of maize, scutellum is considered as cotyledon because it
(A) Protects the embryo
(B) Contains food for the embryo
(C) Absorbs food materials and supplies them to the embryo
(D) Converts itself into a monocot leaf

40. Hypogeal germination is seen in
(A) Tamarind (B) Onion
(C) Cereals (D) Mustard

41. Epigeal germination is seen in
(A) Sunflower (B) Paddy
(C) Lentil (D) Wheat

42. Isolation distance for rice is
(A) 100m (B) 200m
(C) 50m (D) 3m

43. Tetrazolium test is also known as

(A) Biochemical test (B) Biological test
(C) Exchange test (D) Both (A) & (B)

44. Rice variety which don't have seed dormancy ?

(A) IR-8 (B) IR-20
(C) IR-50 (D) Both (A) & (C)

45. Seed sampling is done by

(A) Seed analyst (B) Seed certification officer
(C) Seed inspector (D) All the above

46. Seed Act –

(A) 29th Dec, 1966 (B) 29th March, 1966
(C) 29th Oct, 1966 (D) 29th Aug, 1966

47. Pre-drying is a method done at temperature not exceeding __________ with free circulation period of __________

(A) 40°C, 14 Days (B) 40°C, 7 days
(C) 20°C, 14 Days (D) 20°C, 7 days

48. Most popular substrate plant for biopharming

(A) Cumbu (B) Corn
(C) Wheat (D) All the above

49. Effect of pollen on seed coat colour is called

(A) Meta xenia
(B) Genetic engineering
(C) Transgenic
(D) Fertilization

50. Basic input of Agriculture is

(A) Soil (B) Seed
(C) Irrigation (D) Fertilizer

51. NSC stands for

(A) National Seed Corporation
(B) National seed control
(C) National seed commission
(D) None of the above

52. The branch of seed science and technology deals with

(A) Seed production and processing
(B) Seed certification & testing
(C) Seed storage, marketing and distribution
(D) All the above

53. Which of the following is the goal of seed technology?

(A) Rapid Multiplication and Timely supply
(B) Assured quality of seed
(C) Reasonable price
(D) All the above

54. Botanically, seed is defined as a mature, fertilised, and ripened ovule containing
(A) Viable embryo
(B) Stored food material – cotyledon and endosperm
(C) Protective covering – seed coat
(D) All of the above

55. Any plant part which can be used for commercial multiplication of crop is called as
(A) Seed technology (B) Plant breeding
(C) Seed morphology (D) Seed production

56. National seed corporation starts functioning from
(A) June 1961 (B) July 1963
(C) July 1967 (D) June 1961

57. The methods of science dealing with improving the physical and genetical characteristics of the seed is called as
(A) Seed technology (B) Plant breeding
(C) Genetic engineering (D) Seed production

58. A genotype which is released or notified by CVRC or SVRC is called as
(A) Hybrid (B) Variety
(C) Hybridisation (D) All the above

59. In India, Seed labeling is
(A) Compulsory (B) Tentative
(C) Voluntary (D) Statuary

60. Seed act came into force throughout the country on
(A) 29th Dec, 1969 (B) 2nd Oct, 1969
(C) 30th Dec ,1983 (D) None of the above

61. The main features of Seed act 1966 is/are
(A) Applicability (B) Sanctioning legislation
(C) Regulatory legislation (D) All the above

62. Which class of seed is known as Initial seed?
(A) Nucleus Seed (B) Breeder seed
(C) Foundation seed (D) Certified seed

63. Internal cork of Apple/browning of Cauliflower is deficiency symptom of
(A) Mo (B) B
(C) Cu (D) Zn

64. Whiptail of crucifers is due to deficiency of
(A) Mo (B) Cu
(C) Zn (D) P

65. Which of the following is present in the core of chlorophyll molecule
(A) Fe (B) Mg
(C) K (D) Mn

66. Crossing between two inbred line is called as
(A) Single cross Hybrid (B) Double cross Hybrid
(C) Three cross hybrid (D) Top cross hybrid

67. When F1 is crossed with third inbred line is called as
(A) Single cross hybrid (B) Double cross Hybrid
(C) Three cross hybrid (D) Top cross hybrid

68. Crossing between two hybrids is known as
(A) Single cross hybrid (B) Double cross Hybrid
(C) Three cross hybrid (D) Top cross hybrid

69. The test required for granting the protection of a new variety under PPV&FRA 2001 is
(A) DUS (B) VCU
(C) GOT (D) All the above

70. A seed lot will be eligible for extension of validity period
(A) For Initial nine months Only
(B) For one year only
(C) For two years only
(D) As long as it conforms to the prescribed standards

71. Multiplication of seed germination percentage with seedling dry weight on the day of final count is known as
(A) Germination Index (B) Rate of seed germination
(C) Vigour Index (D) Germination Energy

72. High seed leachate conductivity is indicative of
(A) High seed vigour (B) Low seed vigour
(C) Purity of seed (D) Viability of seed

73. An orthodox seed can
(A) Tolerate dehydration and freezing
(B) Tolerate dehydration but not freezing
(C) Tolerate freezing but not desiccation
(D) None of the above

74. High quality seed are the results of good agricultural practices, which include
(A) Proper maintenance of genetic purity
(B) Good growing condition
(C) Proper timing and method of harvesting
(D) All the above

75. Which of the following practices will reduce seed viability?
(A) Moisture content above 13% in seed
(B) Poor storage condition
(C) Excessive landing of seed
(D) All the above

76. Seed processing can improve
(A) Physical quality (B) Genetic quality
(C) Moisture (D) All the above

77. Plant breeders are interested in Apomixis because it
(A) Enlarges genetic variability
(B) Improves seed set
(C) Yield clone progeny
(D) Shortens the life cycle

78. Hidden infestation in seeds can be established by
(A) Roentgenography (B) Hiltner test
(C) Standard germination test (D) TZ test

79. Section 8 A of the seed act, 1966 pertains to
(A) The Appellate authority (B) The certification agency
(C) Central seed committee (D) The central seed certification board

80. Directorate of seed research is located at
(A) New Delhi (B) Kalimpong
(C) Mau bhanjan (D) Shimla

81. Which of the following private seed multinational companies have the tag line "Breeding Excellence" ?
(A) Nuziveedu Seeds (B) RASI Seeds
(C) Monsanto Seeds (D) Mahyco Seeds

82. Plant breeder's and farmer's rights are conferred as per following act
(A) Seeds Act
(B) Copyrights Act
(C) Geographical Indication of Goods Act
(D) PPV & FR Act

83. GEAC (Genetic Engineering Appraisal Committee) is an apex body, functioning under
(A) Ministry of Agriculture
(B) Ministry of Science and Technology
(C) Ministry of Environment and Forests
(D) Department of Biotechnology

84. Specific gravity separator separates the seeds which are primarily
(A) Similar in size and weight but different shape
(B) Similar in texture but different shape
(C) Similar in shape and size but different weight
(D) Similar in colour but different weight

85. The seed sample sent by a seed control/seed law enforcement officer for determining whether it meets the minimum seed certification standards or not, is referred to as
(A) Service sample (B) Certification sample
(C) Official sample (D) Personal sample

86. Provisions under Section-5 of the Seeds Act, 1966 pertains to
(A) Power to notify kinds or varieties of seeds
(B) Powers to specify minimum limits of germination and purity
(C) Grant of certificate by certification agency
(D) Revocation of certificate

87. Consider the following statements regarding sampling intensity
(i) For seed lot having more than 6 and less than 30 containers, sampling shall be done in minimum of 10 containers or at least one in every five containers, which ever is greater.
(ii) Seeds held in small containers such as tins or packets, a 1000 kg weight of seed are taken as the basi unit
(iii) For seed lot up to 500 kg, a minimum of 5 primary samples must be taken.

Which of the above statements is true?
(A) (i) only (B) (ii) only
(C) (iii) only (D) (i) and (iii)

88. Which among the following statements is false with regard to sample reduction in the laboratory?
(A) Boerner divider is suitable for all kinds of seeds, except for extremely chaffy seeds
(B) Random cup method is not suitable for crops which are extremely chaffy and which bounce or roll
(C) Spoon method is used only for samples of single small-seeded species
(D) Hand halving method is followed for genera which are not chaffy

89. During purity analysis, what should be the minimum number of decimal places recorded during weighing of purity components if the weight of working sample is 7 grams?
(A) 0 (B) 1
(C) 2 (D) 3

90. What is the specification for the size and colour of breeder seed tag in India?
(A) Size: 15 cm × 7.5 cm, Colour: Golden yellow
(B) Size: 12 cm × 6 cm, Colour: Golden yellow
(C) Size: 15 cm × 7.5 cm, Colour: White
(D) Size: 12 cm × 8 cm, Colour: Golden yellow

91. Knee-bent symptom is an essential feature for classification as a normal seedling in germination test of
(A) Soybean (B) Onion
(C) Chickpea (D) All of the above

92. Which one of the following is the minimum limit of germination percentage for certified seed of chickpea as per IMSCS?

(A) 75 (B) 80
(C) 85 (D) 90

93. The maximum weight of a seed lot in case of groundnut crop as per IMSCS should not exceed beyond

(A) 10,000 kg (B) 20,000 kg
(C) 30,000 kg (D) 40,000 kg

94. The determination of seed moisture content through Low Constant Temperature Oven method requires drying temperature and duration of exposure (hours) at

(A) 103±2oC for 24±1 hours (B) 103±2oC for 17±1 hours
(C) 130±2oC for 17±1 hours (D) 130±2oC for 1±4 hours

95. Under which of the following circumstances, the germination test shall be repeated using an appropriate method?

(A) No dormancy is suspected
(B) Seedling showing primary infection
(C) The germination range does not comply to the norms of tolerances
(D) No error in test condition

96. Match the following

Hybrid crop	**Method**
(a) Maize	1. Temperature based sex expression
(b) Castor	2. Detasseling
(c) Chilli	3. Use of gynoecious line
(d) Ribbed gourd	4. Emasculation and dusting

(A) 4, 3, 2, 1 (B) 2, 4, 3, 1
(C) 3, 4, 1, 2 (D) 2, 1, 4, 3

97. Match the following

Crop	**Harvest period**
(a) Redgram	1. 85 - 90 DAS
(b) Sorghum	2. 150 - 180 DAS
(C) Bhendi	3. 75 - 80 DAS
(d) Safflower	4. 35 DAA

(A) 4, 3, 1, 2 (B) 4, 2, 3, 1
(C) 3, 1, 4, 2 (D) 2, 3, 4, 1

98. Match the following

Hybrid	**Year**
(a) First maize hybrid	1. 1965
(b) First cotton hybrid	2. 1964
(c) First sorghum hybrid	3. 1971
(d) First pearl millet hybrid	4. 1961

(A) 4, 3, 2, 1 (B) 4, 2, 3, 1
(C) 3, 4, 1, 2 (D) 4, 3, 2, 1

99. Match the following

(a) ESTA — 1. Switzerland
(b) ISTA — 2. Denmark
(c) AOSA — 3. Canada
(d) CSAAC — 4. USA

(A) 4, 3, 2, 1 (B) 2, 4, 3, 1
(C) 2, 1, 4, 3 (D) 2, 3, 4, 1

100. Match the following

(a) ICAR — 1. 1919
(b) AOSA — 2. 1969
(c) SFCI — 3. 1929
(d) ICIA — 4. 1908

(A) 4, 1, 3, 2 (B) 3, 4, 2, 1
(C) 1, 3, 4, 2 (D) 2, 3, 4, 1

Answer Keys

Q.No.	Ans.	Q.No.	Ans.	Q.No.	Ans.	Q.No.	Ans.	Q.No.	Ans.
1	B	21	A	41	C	61	D	81	B
2	A	22	A	42	A	62	B	82	D
3	C	23	A	43	D	63	B	83	C
4	A	24	D	44	D	64	A	84	C
5	B	25	B	45	D	65	B	85	C
6	B	26	D	46	A	66	A	86	A
7	A	27	A	47	B	67	C	87	C
8	D	28	D	48	B	68	B	88	A
9	C	29	A	49	A	69	A	89	D
10	D	30	A	50	B	70	D	90	A
11	A	31	C	51	A	71	C	91	B
12	C	32	B	52	D	72	B	92	C
13	B	33	B	53	D	73	A	93	B
14	B	34	B	54	D	74	D	94	B
15	B	35	D	55	D	75	D	95	B
16	B	36	A	56	B	76	A	96	D
17	C	37	D	57	A	77	A	97	C
18	C	38	B	58	B	78	A	98	A
19	A	39	C	59	A	79	D	99	C
20	B	40	C	60	B	80	C	100	B

MODEL TEST 20

1. **The characteristic indicated by the blackened figures is probably:**
   (A) Dominant (B) Recessive
   (C) Non-dominant (D) Sex-linked recessive
2. **What are the genotypes of the parents?**
   (A) Both are homozygous dominant.
   (B) Both are heterozygous dominant.
   (C) Both are homozygous recessive.
   (D) The male is homozygous dominant; the female is homozygous recessive.
3. **If one parent has type A blood and the other parent has type B blood, what blood type will the offspring denoted by the white square and circle have?**
   (A) Type A (B) Type B
   (C) Type AB (D) Type O
4. **Mitotic cell division results in two cells that have:**
   (A) n chromosomes and are genetically identical
   (B) n chromosomes and are genetically different.
   (C) 2n chromosomes and are genetically identical
   (D) 2n chromosomes and are genetically different.
5. **In tobacco, if the diploid number of chromosomes is 48, how many chromosomes will be found in a pollen grain?**
   (A) 96 (B) 48
   (C) 24 (D) 12
6. **Which of the following are two nuclei inside the embryo sac**
   (A) Egg cell and secondary nucleus
   (B) Micropile and Egg cell
   (C) Stamen and Stigma
   (D) Anther and Style
7. **Which one of the following is the maximum standard permitted for other crop seeds for foundation seed of maize inbreds?**
   (A) 20/kg (B) 10/kg
   (C) 5/kg (D) None
8. **What do you mean by halopriming?**
   (A) Soaking in water (B) Soaking in osmotic solution
   (C) Soaking in salt solution (D) Soaking in biocontrol agents

9. At what temperature, the seeds possessing an initial moisture of 18-30% should be dried?

(A) 110°F (43.3°C) (B) 100°F (42.2°C)
(C) 90°F (42.2°C) (D) 85°F (30.0°C)

10. While drying, normally the moisture content of seeds is redirected to ___________

(A) 14-16% (B) 13-15%
(C) 12-14% (D) 10-12%

11. Which one of the following is not relevant for land selection for seed production?

(A) Moderate water facilities (B) Good fertile soil
(C) Neutral pH (D) Good drainage facilities

12. Which one of the following crop can be harvested in single harvest?

(A) Cotton (B) Tomato
(C) Rice (D) Chillies

13. What is the symptom of molybdenum deficiency in crucifers?

(A) Black rot (B) Whip tail
(C) Blossom end rot (D) Marsh spot

14. When the seeds are exposed to high temperature at the time of development and maturation, the hard seeds are developed. In which crop, this is noticed

(A) Peas (B) Greengram
(C) Cluster bean (D) Blackgram

15. Apple possess ___________ dormancy

(A) Seed coat (B) Chemical
(C) Secondary (D) Embryo

16. Gossypol, a toxic substance in cotton is normally contained in :

(A) Stem (B) Leaves
(C) Fruit wall (D) Seed

17. Which one of the following is sprayed to pulses for the production of bolder seeds?

(A) 0.5% DAP (B) 1% DAP
(C) 1.5% DAP (D) 2% DAP

18. Which one of the following crop can be harvested in multi picking?

(A) Rice (B) Cotton
(C) Sorghum (D) Blackgram

19. ______ is experienced with dimpled seed when it is exposed to high temperature and humidity at the time of maturity.

(A) Peas (B) Greengram
(C) Cluster bean (D) Blackgram

20. Which of the following statement is incorrect?
(A) Auxins are the most important plant hormone.
(B) Auxins are produced at the region of elongation.
(C) Indoleacetic Acid (IAA) is a principal auxin.
(D) Auxins are also important in regulating the fall of leaves and fruits

21. For conducting osmotic stress, which chemical is used
(A) Potassium chloride (B) Sodium chloride
(C) Inositol (D) Mannitol

22. Genetic purity for certified seed is __________ %
(A) 100 (B) 99.9
(C) 99.5 (D) 99

23. Which of the following contains reserve food material
(A) Stamen (B) Stigma
(C) Endosperm (D) Anther

24. Distance between stigma and anthers in same flower,
(A) Reduces chances of self pollination
(B) Enhances chances of self pollination
(C) Reduces chances of cross pollination
(D) Enhances chance of cross pollination

25. What is the temperature and duration of oven drying for moisture estimation using low constant temperature method?
(A) $101 \pm 2°C$ for $17 \pm 1h$ (B) $102 \pm 2°C$ for $17 \pm 1h$
(C) $104 \pm 2°C$ for $17 \pm 1h$ (D) $103 \pm 2°C$ for $17 \pm 1h$

26. _______ is the minimum standard for germination of certified seed of cotton hybrid.
(A) 60% (B) 65%
(C) 70% (D) 75%

27. The toxic substance gossypol is found in
(A) Castor (B) Cotton
(C) Groundnut (D) Rapeseed

28. This is the process by which the plant embryo or baby plant within the seed grows and the baby seedling comes forth from the seed.
(A) Pollination (B) Fertilziation
(C) Germination (D) Anthesis

29. First part to come out of seed is the
(A) Radicle (B) Plumule
(C) Stem (D) Cotyledon

30. In which one of the following country, DLF-Trifolium is located?
(A) Denmark (B) Switzerland
(C) Germany (D) Japan

31. What is the SMR for sesamum?

(A) 1:49 (B) 1:150
(C) 1:120 (D) 1:200

32. Which of the following is not true about synthetic seeds?

(A) can be stored for a year without the loss of valuables
(B) easy to handle
(C) can be directly ssown in the soil like natural seeds
(D) need hardening in the greenhouse

33. Working sample is the reduced representative of _________.

(A) Primary sample (B) Composite sample
(B) Submitted sample (D) Working sample

34. In which container, the seed sample should be packed for testing moisture content at STL

(A) Moisture resistant container
(B) Moisture proof container
(C) Moisture vapour proof container
(D) Moisture pervious container

35. Which one of the following is not relevant for seed village with respect to its disadvantages?

(A) Lack of co-operation among farmers
(B) Engage trained labourers
(C) Outbreak of diseases in whole area
(D) Operational difficulties at a particular

36. Seed test that provides information on weed seed and inert material is called

(A) Clean seed test (B) Purity test
(C) Wholesomeness test (D) Viability test

37. Which one of the following is optimum pH of TZ solution for proper staining of living part of seed

(A) 4 (B) < 4
(C) 7 (D) > 7

38. In which state, Ranebennur, well known place for hybrid and flower seed production is located?

(A) Tamilnadu (B) Karnataka
(C) Andhra Pradesh (D) Telungana

39. Which one of the following is a designated inseparable other crop seed for barley?

(A) Wheat (B) Cumbu
(C) Rye (D) Sorghum

40. Which one of the following part of a seed act as a physical barrier

(A) Cotyledon (B) Embryo
(C) Scutellum (D) Pericarp

41. With an increase in seed size upon germination, dry mass of seeds
(A) Increases (B) Decreases
(C) Vanishes (D) Grows proportionally with seed

42. Which of the following is (are) associated with germination?
(A) Increase in water content (B) Resumption of catabolism
(C) Intake of CO_2 (D) DNA

43. Which type of vigour test is GADA test?
(A) Physical test (B) Performance test
(C) Biochemical test (D) Stress test

44. Who has contemplated the TZ test for evaluation of vigour of a seed lot?
(A) Hiltner (B) Moore
(C) Abdul-Baki and Anderson (D) Mathews and Bradnock

45. What is the seed sample size for grow out test?
(A) 200 seeds (B) 400 seeds
(C) 600 seeds (D) 800 seeds

46. Consider the following statements and choose the correct option.
1. The genetic constitution of a plant is unaffected in vegetative propagation
2. Rhizome in ginger serves as an organ of vegetative reproduction
3. Totipotency of cells enables us to micropropagate plants

(A) Statement 2 alone is true
(B) Statement 1 and 2 alone are true
(C) Statements 2 and 3 alone are true
(D) All the three statements 1, 2 and 3are true

47. Vegetative propagation through leaves takes place in
(A) Bryophyllum (B) Ginger
(C) Potato (D) Tapioca

48. How many field inspections are done for cluster bean
(A) 2 (B) 3
(C) 4 (D) 5

49. In which part of the seed, the amylase secretes?
(A) Pericarp (B) Embryonic axis
(C) Aleurone layer (D) Cotyledon

50. A gas required for germination of pea seed is
(A) Nitrogen (B) Oxygen
(C) Hydrogen (D) Water vapours

51. The smallest and the lightest seed is that of
(A) Lemna (B) Morus
(C) Orchis (D) Wolffia

52. The largest and heaviest seed is that of

(A) Lodoicea (B) Coconut

(C) Rafflesia (D) Mango

53. The seed bearing plants belong to the group of

(A) Pteridophyta (B) Bryophyta

(C) Vascular cryptogams (D) Phanerogams

54. In which type of crops, synthetic varieties are breeded

(A) Cross pollinated crop

(B) Self pollinated crop

(C) Often cross pollinated crop

(D) Asexually propagated crop

55. What is the seed renewal period of composite variety?

(A) 2 – 3 years (B) 3 – 4 years

(C) 1 – 2 years (D) 1 year

56. Which one of the following is seed upgrading machine?

(A) Debearder (B) Destoner

(C) Pre-conditioning/scalper (D) Gravity separator

57. The priming agent, which does not penetrate the cell wall is

A.PEG (B) Water

C.Mannitol (D) Halogens

58. A problem weed, whose seed once mixed with crop seed is extremely difficult to separate is called as

(A) Non parasitic weed (B) Objectionable weed

(C) Parasitic weed (D) Aquatic weed

59. The field carry over storage pest in pulses is

(A) Pulse pod bug (B) Pulse pod fly

(C) Pulse pod wasp (D) Pulse beetle

60. The maximum temperature for safe drying of orthodox seeds is

(A) 40°C (B) 41°C

(C) 42°C (D) 43°C

61. Find the crop species that is mismatched with its location of inhibitor causing dormancy

Coriandrum sativum	Hull
Beta vulgaris	Pericarp
Helianthus annus	Pericarp, testa
Hordeum vulgare	Hull

(A) *Coriandrum sativum* (B) *Beta vulgaris*

(C) *Helianthus annus* (D) *Hordeum vulgare*

62. When the pathogen is loosely mixed with seed in the form of sclerotia, galls etc are called

(A) Infection (B) Infestation

(C) Contamination (D) Concomitant contamination

63. ________ is a statutory body under the Environment Protection Act 1986 of the Ministry of Environment & Forests (MoEF) for approving and regulating GMO crops

(A) Genetic Engineering Appraisal Committee
(B) GOI
(C) Biotechnolgy regulatory authority
(D) All of these

64. India is a ________ Largest contributor of GMO crops.

(A) Second (B) Third
(C) Fourth (D) Fifth

65. GM traits is higher for

(A) Insect resistance (B) Herbicide resistance
(C) Delays fruit ripening (D) None of these

66. Field inspection fee per acre for hybrids is

(A) Rs 40/- (B) Rs 65/-
(C) Rs 130/- (D) Rs 60/-

67. Seed bill was introduced in

(A) 2001 (B) 2004
(C) 2006 (D) 1983

68. The period of protection for notified varieties in PPV & FR Act is

(A) 9 years (B) 15 years
(C) 18 years (D) 12 years

69. Choose the correct answer

(i) There are 25 sections in Seed Act 1966
(ii) PPV & FR Act was enacted in the year 2001
(iii) National Seed Policy 2002
(iv) New policy on seed development 1988

(A) (iii) and (ii) are correct (B) (i) and (ii) are correct
(C) (i) and (iv) are correct (D) (i), (ii), (iii) and (iv) are correct

70. Seed certification was started in India since

(A) 1986 (B) 1963
(C) 1983 (D) 1996

71. Minimum number of field counts to be taken in an area of 5 ha seed field

(A) 5 (B) 6
(C) 7 (D) 8

72. The permeability of seed to the water is decreased is also decreased due to soil

(A) Bacteria (B) Fungi
(C) Composition (D) Both (A) and (B)

73. In Coffee and yucca seed, carbohydrate is stored in

(A) Radicle/hypocotyls (B) Cotyledon
(C) Embryonic tissues (D) Perisperm

74. Osmotic solution used for demonstration of cell elongation and radical protrusion

(A) PEP (B) PEG and mannitol
(C) Glycin betaine (D) Sorbitol

75. Protein accumulate during Arabidopsis seed germination

(A) Gluten (B) Actin -7
(C) Proline (D) Zeatin

76. Number of seeds to be tested for assessing germination percentage as per ISTA is

(A) 100 (B) 400
(C) 200 (D) 240

77. The light intensity which favours seed germination

(A) 700-780mm (B) 600-660mm
(C) 310-450mm (D) 280-300mm

78. Which of the following is not the necessary factors for germination?

(A) Ability of seed to germinate (B) Moisture
(C) Temperature (D) None

79. National seed programme was started in

(A) 1974 (B) 1988
(C) 2002 (D) 1969

80. Seed multiplication ratio of brinjal is

(A) 1:400 (B) 1:450
(C) 1:240 (D) 1:171

81. How many primary samples are to be drawn from a seed lot having 6 to 30 containers?

(A) Atleast one in every 3 containers
(B) Atleast one in every 7 containers
(C) Atleast one in every 5 containers
(D) Atleast 5 primary samples

82. Which one of the principle is followed for mixing and dividing of seeds using Boerner divider?

(A) Centrifugal force (B) Electrical force
(C) Magnetic force (D) Gravitational force

83. Which category of seed yields the milky substance when it is pressed at the end of germination test?

(A) Hard seed (B) Imbibed seed
(C) Dead seed (D) Fresh ungerminated seed

84. What is the percentage of sterility in GMS?

(A) 100% (B) 50:50%
(C) 25:75% (D) 75:25%

85. In which one of the family, the protogyny mechanism exist?

(A) Malvaceae (B) Convolvulaceae
(C) Magnoliaceae (D) Leguminosae

86. Magnesium is required for
(A) DNA replication
(B) RNA synthesis
(C) Stomatal movements
(D) Synthesis of chlorophyll

87. Potassium is employed in
(A) Chlorophyll synthesis
(B) Protein synthesis
(C) Cementing synthesis
(D) Opening and closing stomata

88. Calcium is a component of
(A) Middle lamella
(B) Primary walls
(C) Secondary walls
(D) Chlorophyll

89. Both nitrogen and sulphur and required by plants for
(A) Chlorophyll synthesis
(B) Enzyme
(C) Cell wall
(D) Stomatal movements

90. Micronutrients mainly function as
(A) Osmotic constituents of cell sap
(B) Components of important biochemicals
(C) Cofactors of enzymes
(D) Constituents of chlorophyll

91. Which of the following terms best describes the statement "the angle with the horizontal plane at which the seed material will stand when piled" ?
(A) Resilience
(B) Angle of repose
(C) Free settling velocity
(D) Angle of friction

92. Which of the following statements is false?
(A) The validity period shall be nine months from the date of test at the time of initial certification
(B) The extension of validity period of certified seed shall be for a period of six months provided it confirms to the prescribed standards
(C) The certification agency shall preserve the guard samples of seed lot for a period of four years in case of rejected seed lot from the date of communication of rejection
(D) Under intercropping for production of certified seed class, the crops selected should belong to same genus and preferably with different maturity

93. Which of the following statements is false with respect to Field Counts conducted during Seed Crop Inspection?
(A) It is necessary to take a minimum of 3 counts up to 2 ha of seed production and an additional count for every additional 2 ha
(B) For a seed production area of above 6 to 8 ha, the minimum number of counts to be made is 8
(C) Number of plants/heads per count is sunflower shall be 100
(D) Number of plants/head per count in sorghum shall be 1000

94. The "Hub and Spoke" model put forth by Prof. M.S. Swaminathan relates to

(A) Agro processing linkages
(B) Human resource management
(C) Village community seed banks
(D) Permafrost seed storage

95. As per Indian Minimum Seed Certification Standards, the minimum permissible plants/heads affected by loose smut in wheat Foundation Class of Seeds shall be

(A) 0.05% (B) 0.10%
(C) 0.50% (D) 1.0%

96. Match the following

Cole crops	Spacing
(a) Cabbage	1. 30 × 20 cm
(b) Carrot	2. 45 × 45 cm
(c) Radish	3. 40 × 30 cm
(d) Beet root	4. 60 × 45 cm

(A) 4, 3, 2, 1 (B) 4, 2, 3, 1
(C) 3, 1, 4, 2 (D) 3, 2, 4, 1

97. Match the following

Cole crops	Seed rate
(a) Cauliflower	1. 3 - 4 kg/ha
(b) Turnip	2. 600 - 750 g/ha
(c) Beet root	3. 9 - 12 kg/ha
(d) Radish	4. 4 - 5 kg/ha

(A) 4, 1, 3, 2 (B) 2, 4, 3, 1
(C) 2, 1, 4, 3 (D) 2, 4, 1, 3

98. Match the following

Forage & fodder crop	Isolation distance (CS)
(a) Cluster bean	1. 10 m
(b) Fodder cowpea	2. 5 m
(c) Lucerne	3. 25 m
(d) Napier - bajra	4. 100 m

(A) 4, 3, 1, 2 (B) 4, 2, 3, 1
(C) 3, 1, 4, 2 (D) 2, 3, 4, 1

99. Match the following

(a) Mango	1. Recalcitrant seed
(b) Citrus	2. Orthodox seed
(c) Guava	3. Aggregate fruit
(d) Jack	4. Berry

(A) 4, 1, 3, 2 (B) 4, 3, 2, 1
(C) 2, 1, 4, 3 (D) 3, 2, 1, 4

100. Match the following

	Hybrid crop		**Isolation distance**
(a)	Date palm	1.	Drupe
(b)	Avacado	2.	Berry
(c)	Persimmon	3.	Recalcitrant seed
(d)	West Indian cherry	4.	Orthodox seed

(A) 4, 3, 2, 1 (B) 1, 4, 2, 3
(C) 2, 1, 4, 3 (D) 3, 2, 4, 1

Answer Keys

Q.No.	Ans.	Q.No.	Ans.	Q.No.	Ans.	Q.No.	Ans.	Q.No.	Ans.
1	A	21	D	41	A	61	A	81	A
2	B	22	D	42	A	62	D	82	D
3	D	23	C	43	C	63	A	83	C
4	C	24	A	44	B	64	C	84	B
5	C	25	D	45	B	65	B	85	C
6	A	26	C	46	D	66	C	86	D
7	C	27	B	47	A	67	B	87	D
8	C	28	C	48	A	68	B	88	A
9	C	29	A	49	C	69	D	89	B
10	D	30	A	50	B	70	A	90	C
11	A	31	D	51	C	71	C	91	B
12	C	32	B	52	A	72	D	92	D
13	B	33	B	53	D	73	D	93	A
14	D	34	C	54	A	74	B	94	C
15	D	35	B	55	B	75	B	95	B
16	D	36	B	56	D	76	B	96	A
17	A	37	C	57	A	77	A	97	C
18	D	38	B	58	B	78	D	98	D
19	C	39	A	59	D	79	A	99	C
20	B	40	D	60	B	80	B	100	A

■ ■ ■

More Books

Titles	Authors
— Basic Concepts of Plant Science	*S.K. Bangarwa*
— Competitive Nematology	*N.G. Ravichandra*
— Chickpea (Gram) Status and Cultivation Technology	*S.S. Shekhawat*
— Forest Tree Seeds: Handbook	*S.S. Sagwal*
— Fundamentals of Agriculture (Vol. 1-2) (Set)	*R.L. Arya*
— Glimpse on General Agriculture (for ICAR-JRF, SRF, NET AND ASRB PRELIMS)	*P. Laxman Rao*
— Hand Book of Seed Industry Prospects and Its Costing	*Avinash Ramdeo*
— Handbook of Seed Science and Technology	*A.S. Basra*
— Identification of Crop and Weed Seeds	*A.F. Musil*
— Indira's Objective Agricultural Biotechnology *2nd Edition*: MCQ`s for Agricultural Competitive Exams	*R.L. Arya*
— Krishi Jagat: JET Evam Krishi Paryavekshak Pariksha Upyogi	*M.L. Jat*
— MCQ's In Microbiology: Advanced	*Balaram Mohapatra*
— MCQ's In Plant Breeding Biotechnology and Seed Science	*K. Vanangamudi*
— Medicinal Seeds and Plants	*K. Vanangamudi*
— Moth Bean In India	*D. Kumar*
— Multiple Choice Questions (MCQ`s) in Food Technology	*P.M. Kotecha*
— Objective Seed Science and Technology	*K. Vanangamudi*
— Objective Agribusiness Management *3rd Edition*	*S.R. Panigrahy*
— Objective Plant Physiology *3rd Revised Edition:* MCQ's In Plant Physiology	*P. Dwivedi*
— Organic Seed : Traditional Varieties and Technologies	*K. Vanangamudi*
— Principles of Seed Technology	*Dhirendra Khare*
— Principles and Practices of Seed Storage	*O.L. Justice*
— Question Bank : Seed Science and Technology	*K. Vanangamudi*
— Seed Identification Manual	*A.C. Martin*
— Seed Borne Diseases: Ecofriendly Management	*Arun Arya*
— Seed Economics: Commercial Considerations for Enterprise Management In Developing Countries	*S. Kugbei*
— Seed Technology in The Tropics	*D.B. Mackay*
— Seed Technology *2nd Revised & Enlarged Edition*	*Dhirendra Khare*
— Seed Production of Agricultural Crops	*A.F. Kelly*
— Seed Quality Enhancement : Principles and Practices	*K. Vanangamudi*
— Seed Economics: Commercial Considerations for Enterprise Management In Developing Countries	*S. Kugbei*
— Vegetable Crops At A Glance	*S.K. Tyagi*
— Weed Seed Biology	*K. Vanangamudi*